T0192448

Management in the Built Environment

Series editor

Low Sui Pheng, National University of Singapore, Singapore, Singapore

The aim of this book series is to provide a platform to build and consolidate a rigorous and significant repository of academic, practice and research publications that contribute to further knowledge relating to management in the built environment. Its objectives are to:

(1) Disseminate new and contemporary knowledge relating to research and practice in the built environment
(2) Promote synergy across different research and practice domains in the built environment and
(3) Advance cutting-edge research and best practice in the built environment

The scope of this book series is not limited to "management" issues per se because this then begs the question of what exactly are we managing in the built environment. While the primary focus is on management issues in the building and construction industry, its scope has been extended upstream to the design management phase and downstream to the post-occupancy facilities management phase. Management in the built environment also involves other closely allied disciplines in the areas of economics, environment, legal and technology. Hence, the starting point of this book series lies with project management, extends into construction and ends with facilities management. In between this spectrum, there are also other management-related issues that are allied with or relevant to the built environment. These can include, for example cost management, disaster management, contract management and management of technology.

This book series serves to engage and encourage the generation of new knowledge in these areas and to offer a publishing platform within which different strands of management in the built environment can be positioned to promote synergistic collaboration at their interfaces. This book series also provides a platform for other authors to benchmark their thoughts to identify innovative ideas that they can further build on to further advance cutting-edge research and best practice in the built environment.

Editorial Advisory Board:

If you are interested in submitting a proposal for this series, please kindly contact the Series Editor or the Publishing Editor at Springer:

Low Sui Pheng (bdglowsp@nus.edu.sg) or
Ramesh Premnath (Ramesh.premnath@springer.com)

More information about this series at http://www.springer.com/series/15765

Low Sui Pheng

Project Management for the Built Environment

Study Notes

 Springer

Low Sui Pheng
Department of Building
National University of Singapore
Singapore
Singapore

ISSN 2522-0047   ISSN 2522-0055 (electronic)
Management in the Built Environment
ISBN 978-981-13-4980-5   ISBN 978-981-10-6992-5 (eBook)
https://doi.org/10.1007/978-981-10-6992-5

Printed on acid-free paper

This Springer imprint is published by Springer Nature
The registered company is Springer Nature Singapore Pte Ltd.
The registered company address is: 152 Beach Road, #21-01/04 Gateway East, Singapore 189721, Singapore

Preface

Project management is as old as the history of mankind. Have you ever wonder who built the ancient pyramids of Egypt? Or who managed the building of the Great Wall in China? And closer to home, who managed the building of landmark Singapore projects such as the iconic Marina Bay Sands Integrated Resorts and the world-renowned Changi International Airport? It's the project manager who leads his team of designers and builders.

Projects need not be limited to iconic buildings. Projects are temporary work assignments to create a definitive deliverable, service or environment. All projects have defined scope, deadlines, budget, resource needs and end results that meet client's requirements. Projects are therefore temporary in nature but not the deliverables. Singapore's Garden-By-The-Bay will be around much longer than the time taken to build the project. But there are exceptions. Our National Day parade and its supporting infrastructures take months to plan and build, but the celebrations event last only a few hours. Once the project is completed, the project team disbands and its members move on to other projects. All projects are unique even if the team does the same type of project over and over again. The time it takes, the stakeholders involved and the environment are unique in each project.

What then is project management? Project management is about applying the knowledge, skills, tools and techniques to meet the requirements of the project client. This is accomplished through the five project management processes of initiating, planning, executing, monitoring/controlling and closing. Project management takes into consideration stakeholders with different needs and expectations. For example, while the client may wish to complete his project quickly, workplace safety must never be compromised. The competing demands of a project in terms of time, cost, quality and risks must be properly ascertained and managed.

Most projects start with an idea or concept which needs to be elaborated progressively to flesh out the details. Research complements project management to further refine the initial concept leading to concept clarification. With the concept clarified, a feasibility study is then conducted to establish the viability of the project. This provides the basis to define the project scope for the project manager to commence planning.

The project manager is guided by a body of knowledge in the planning process. This body of knowledge includes ten areas relating to integration, scope, time, cost, quality, human resource, communication, risks, procurement and stakeholder management. These knowledge areas establish for the project, what must be done, when should it be done, how much it will cost, how good should it be, who will do the work, how will information be delivered, what problems may be encountered, what resources must be obtained and how buy-in for the project can be obtained.

These knowledge areas can fit into different industries such as manufacturing, consulting, banking, construction and tourism. However, the approach to practicing project management would be the same regardless of the industry the project manager is in. Nevertheless, he needs to be aware of the different influences on project outcomes from the physical environment, cultural and social environment as well as the international and political environment.

In more complex project environments such as those found in the construction sector, these knowledge areas can be further extended to cover safety concerns, environmental issues, financial and claims management.

To manage the above influences, the project manager needs to possess general management and interpersonal skills. Apart from being able to plan, organize and control, the project manager needs to have good problem-solving and negotiation skills. As a leader, he is expected to motivate and communicate well to positively influence the project organization.

In reality, all of us have in one way or another functioned as accidental project managers without us even realizing it. Project management in its most elementary form is pervasive in many areas of our daily lives. In schools, students manage their project work. They work in unison with their classmates to deliver an academic product that meets prescribed quality standards within a stipulated submission deadline. It's a project when a school moves from existing facilities to a new building. In housing, when we renovate our homes, we are actually project owners who engage the services of the interior designers and contractors for the works. As project owners, we are involved with setting the scope, time, cost and quality targets for our house renovation projects. When an organization migrates from one computer platform to another, it's an info-communication technology project. Designing and prototyping a revolutionary fighter jet plane can be a new defence project. Learning more about effective project management should be our appropriate response to this realization.

In this book, you will learn about the five project management processes relating to: initiating, planning, executing, monitoring/controlling and closing. You will also learn about nine specific project management knowledge areas (PMBOK) relating to: integration, scope, time, cost, quality, human resource, communication, risks and procurement. You will also learn how these project management processes and knowledge areas map over different project life cycles and phases. In the last chapter of this book, you will also appreciate the professional code of conduct and ethical practices that all professional project managers subscribe to. This book does not cover stakeholder management, safety and environment as well as financial and claims management because their related issues would already be discussed in the above-mentioned knowledge areas. Stakeholder management, for example, could be subsumed under human resource management and communications management.

The global construction market by 2020 is estimated to be worth some US$10.3 trillion at constant 2010 prices. The worldwide opportunities for construction project managers are clearly very attractive. It has also been said that the construction industry is by far one of the most complex industries. By extension, if you have had experience as a construction project manager, the likelihood is that you can readily transfer your professional skill sets across sectorial boundaries to work in other industries both locally and globally.

Singapore, Singapore   Prof. Low Sui Pheng
                                         Director, Centre for Project Management
                                                          and Construction Law

Contents

Chapter 1
Introduction to Project Management

1.1 Introduction

a. What is project management?

Projects are not new. Since young, all of you have performed the role of the accidental project manager even without you taking any formal courses in project management. These projects might include putting together a house model or constructing a small playground for a school in another developing country as part of an overseas community involvement project. Building projects have been completed since time immemorial; from the Great Wall of China to the ancient pyramids of Egypt. So what exactly does the modern-day project manager do? First, the project manager needs to understand the project requirements from the client's perspective. He then uses his know-how, skillsets, tools and techniques to meet these requirements that are often related to time, cost and quality targets. In achieving these targets, he moves through the five project management processes of initiating, planning, executing, monitoring and controlling as well as closing. To do this job well, it is therefore crucial for you to be familiar with the project management body of knowledge (which is pronounced as PIMBOK).

b. Competing project demands

There are many stakeholders involved with a project. This is especially true for projects that are large and complex in nature. In such projects, there are invariably different needs and expectations among the stakeholders. In most projects, resources are also limited and the project manager is therefore guided first and foremost by the budget set by the client. The project manager therefore needs to strike a fine balance between the budget set and different stakeholders' demands with respect to project scope, quality, time, and risk appetite. It is therefore important for the project manager to identify what these requirements, expectations and competing demands are as early as possible.

© Springer Nature Singapore Pte Ltd. 2018
L.S. Pheng, *Project Management for the Built Environment*,
Management in the Built Environment,
https://doi.org/10.1007/978-981-10-6992-5_1

c. What others say about project management

There are many literatures written about project management. Some of the key terms mentioned in these literatures include the following: organization, work breakdown, cost estimating, planning, scheduling, purchasing, change management, risk, written charter, leadership, project team, project plan, vendors, communications, and closing phase. Hence, it can be noted that many different terms are used in project management that you should be familiar with at the end of this course. Project management is also applied in different sectors as diverse as construction, manufacturing, banking, hospitality, information, transport and communication as well as oil and gas. Project management also cuts across national boundaries as projects move from the domestic market to an overseas market. Consequently, cultural management should also be considered.

d. What the professional institutions offer

There are many professional institutions worldwide that are related to the practice of project management. These include the International Project Management Association, the Association of Project Management based in the United Kingdom and the Project Management Institute based in the United States. While the common focus is on project management, each of these professional institutions deals with the subject matter differently. The International Project Management Association includes legal issues, interpersonal issues, relationship management, marketing, product functionality and strategic alignment in its treatment of project management. The Association of Project Management identifies the five stages of project management practice to include opportunity identification, design and development, implementation, hand-over and post-project evaluation. The Project Management Institute sets out its project management body of knowledge to include the nine knowledge areas of integration management, scope management, time management, cost management, quality management, human resource management, communications management, risk management and procurement management. These nine knowledge areas are commonly pronounced as PIMBOK which you will learn in the course of this module.

1.2 What are Projects?

a. The nature of Projects

Projects can both be small and large. Projects can also take place in different environments such as in construction, banking and education. The environments within which projects can take place are limitless. However, in all cases, projects are normally temporary work endeavors or assignments. They have a start time and an end time. Apart from the start and end deadlines, projects are also constrained by limited resources, pre-defined budgets and specified end results. Upon completion, projects create some agreed upon deliverables. These can be in the form of a service such as the setting up of a call-centre or in the form of a physical entity such as a school building.

b. Temporal nature of projects

Project management involves both the processes as well as the end-product. Project processes are therefore temporary in nature and end upon the delivery or creation of the project deliverable which can last for many years. For example, in the construction of a large shopping mall, the construction process may take between two to three years. Upon completion and with proper maintenance and management, the shopping mall is expected to be functional for many years. On the other hand, a concert to raise funds for charity can take many months to plan. But the concert only lasts a few hours. In both the shopping mall and the concert, the project team is assembled at project inception and upon project completion, team members disband and return to their own organizations to take on other projects.

c. Uniqueness of projects

While the project management body of knowledge is generic in nature, projects on the other hand, are unique. For example, each shopping mall is different. Externally, its location and site conditions are clearly different from another shopping mall elsewhere. Internally, its fire protection system and shop layout are also likely to be different from another shopping mall. The uniqueness of each project can also be extended beyond the construction industry to the service industry where each call-centre created is different and each inventory management system put in place is also different. Each project is different because its operating environment, resources available, stakeholders involved, constraints faced and the duration taken for its completion are different. This is so even if the same project manager or the same stakeholder is involved with the same type of project over and over again. A construction firm may specialize in the erection of school buildings. However, each school is likely to be unique as the firm faces a different team of stakeholders, site location and operational constraints altogether. Nevertheless, this does not mean that the lessons learned from the completion of each of these school projects cannot be transferred to another new school project so that the same mistakes can be avoided altogether.

d. Progressive elaboration and development

Projects do not just appear out of nowhere. All projects go through a process known as progressive elaboration and development. This starts with what appears to be a good idea or a project concept from someone. However, at this stage, it is not clear if the project concept will work out well in real life or can be achieved realistically. Research or further study is therefore needed to evaluate if the project concept or idea is feasible, viable and achievable. At this stage, project research activities are carried out. With the appropriate research findings and with more facts in hand, the project concept is then further refined or tweaked to meet stakeholder requirements. Further questions may be asked to provide for greater clarity. This stage is known as the project validation or clarification stage. Upon finalization of this stage, feasibility studies are conducted to verify the project specifications. The project feasibility stage sets out to ensure that the time, cost, quality, and other project objectives can be met.

Not all project feasibility stage will end up with meeting objectives that are measured in monetary terms. Some project objectives are spelled out in non-monetary terms such as those associated with contributions to society or for a specific charitable cause which is non-profit in nature. After the feasibility studies are completed and acceptable to the stakeholders, the scope of the project is then defined. The project scope so defined provides the guidance to the project team on what is in and what is outside the scope of the project. The project team is then able to develop and work out the details of the project plan for implementation. The project plan can be further elaborated through a work breakdown structure that consists of the components needed for the successful completion of the project. These components can in turn be described by their respective activities. Correspondingly, the project plan is elaborated and developed further based on the project management body of knowledge as well as the project management processes. In summary, project elaboration and development provide the refinements that project components (and their respective activities) pass through to reach their final state as desired by the project client.

1.3   Projects, Events and Operations

a. Defining projects

At times, it is difficult to draw a fine line between a project and an event. A project usually has some defined milestones and goes through a project life cycle to eventually deliver the outcome. In the case of construction, these milestones can include the completion of the foundation works, super-structural works, architectural works, as well as the utilities and services. While there are deadlines for the completion of each milestone as well as the overall project, these milestones may shift over time and can be completed earlier or later than planned or can be postponed altogether. The timely completion of each milestone is also crucial for the next milestone to commence although some works can be carried out concurrently. In the case of construction, super-structural works generally cannot take place until the foundation works have been completed. However, architectural works can be carried out at the same time when utilities and services are being installed. Projects can also take place outside of the construction industry and in any businesses. Projects can include the design and production of a new submarine for the navy, implementing a new computer operating system within a ministry, moving the operations of a business from one city to another, and running the local elections for a political party.

b. What are events?

On the other hand, an event does not have defined milestones that go through a life cycle. However, some key dates can be indicated when some critical activities are expected to be completed. An example can be an event for a travel fair to be carried out in a major convention centre. For the travel fair to materialize, a venue must first be secured. There must be plans for the construction of the booths for the exhibitors, security, provisions for crowd control, food and beverages outlets, and so on. These can take place at the same time or at different time and some delays may be tolerated. However, the start date for the event cannot be changed. All these works must be completed for the travel fair to be opened on a date and time that have been communicated to members of the public.

Thereafter, year after year, the same travel fair event may be carried out either in the same venue or another venue. The resources and the preparatory works needed to run this travel fair event are also likely to be the same. The same people may also run this travel fair event from year to year. Unless the event is large, complex, and prestigious and of national significance, there is normally no practice, rehearsal or trial runs for events. One example of such an event where national prestige is of utmost importance is the National Day or Independence Day parade where performers get together to practice and rehearse before the actual event. In the case of events, success is not measured by the processes leading to the actual event launched. Success is indicated by the actual event itself and in communicating to people that the event is to take place.

c. What are operations?

Projects for the same organization can be unique. This is because the activities that are carried out for these projects are different even within the same organization. Projects have different completion deadlines and may be influenced by different constraints and operating environments. When projects are completed, these are handed over to the clients who then manage, maintain and operate the completed projects. For example, the completion of a shopping mall project leads to operations within the mall. For this reason, it is often difficult to separate projects from operations. Both projects and operations require resources in the form of people, equipment, materials and financing.

In this scenario of the construction project and the completed shopping mall, the project manager and facilities manager are the two key persons involved respectively with identifying, designing, executing and managing these resources. For the facilities manager, routine day-to-day operations are required to keep the shopping mall running in an efficient manner to provide service quality not only to the shoppers but also the shop tenants. For a large shopping mall, the facilities manager is often stationed within the mall to ensure that the day-to-day operations are carried out without any hitch. On the other hand, organizations that undertake construction projects can include the property developer, architect, engineer, quantity surveyor and builder. For these organizations, projects are therefore short-term tasks that must be completed within the stipulated project duration. However, for organizations that undertake works for their clients, everything can be a project. The builder is a good case in point and he can be constructing a shopping mall and a factory for two different clients at the same time. Both the shopping mall and the factory are projects in so far as the builder is concerned.

1.4   Why are Projects Created?

a. Strategic issues for projects

Projects are created only after careful consideration of their feasibility in satisfying the strategic needs and goals of the various stakeholders. Projects are often identified based on marketplace conditions and the associated supply and demand situations. The completed projects can serve to contribute to the bottom line of a business, productivity or for a charitable cause. There are many examples why projects are created. A government ministry may find it more productive when all employees work on standardized office applications to better facilitate the seamless transfer of electronic information. For this reason, a project to replace all computers with a common operating system in the ministry is thus created. A property developer may decide to embark on the construction and sale of elder-friendly studio flats to meet the demands of a greying population.

Businesses may also constantly initiate company-wide projects to exploit the use of new technologies. Supermarkets may therefore replace current barcoding with real-time frequency identification technology for more efficient check-outs by customers. Pharmaceutical companies can also embark on research and development projects to discover new drugs for the treatment of certain diseases. The introduction of new laws and regulations can also trigger the creation of new projects to meet demands from businesses. The introduction of new workplace safety and health regulations can therefore spur safety consultants to create new services relating to risk assessment and evaluation projects. Projects are therefore created to meet the strategic goals of organizations and must therefore be well planned for these goals to be achieved.

b. How project management is defined

The magnitude, nature and scale of projects can vary from organization to organization and from industry to industry. Regardless of these potential differences, project management in these organizations and industries can be defined to mean the planning, organizing and controlling of activities to successfully accomplish the project's vision and goals. Projects typically include the project manager and the project team. The project manager is responsible for ensuring that resources are in place for project tasks to be carried out. The project manager then plans, schedules, monitors and controls each of these tasks for successful and timely completion. For large projects, the project manager is aided by his project team members who ensure that the various tasks identified are completed as planned. The practice can differ for small projects where a roaming project manager is responsible for overseeing more than one project on a daily basis.

1.5 Project Management Body of Knowledge

a. What the body of knowledge encompasses

There are nine areas of study in the project management body of knowledge. These are integration management, scope management, time management, cost management, quality management, human resource management, communications management, risk management and procurement management. These nine knowledge areas are underpinned by five project management processes. These processes are initiating, planning, executing, monitoring and controlling, as well as closing. All nine areas in the body of knowledge go through the five processes. For this reason, there can be overlaps between the nine areas in the body of knowledge as a change in one area can correspondingly affect changes in other areas. For example, a change in the scope of a project can lead to corresponding changes in time, cost, quality, human resource, communications, risk, and procurement requirements.

Project co-ordination is therefore necessary to ensure that all changes are synchronized. This co-ordination is facilitated through project integration management. In the construction industry, it is quite common for changes or variations to take place early, midway or towards the end of a project. For example, a small condominium project may be conceived earlier without a swimming pool. Following feedback from the project marketing team and midway through the construction phase, the developer may decide to include a swimming pool to entice buyers. This addition of a swimming pool is bound to have implications for time, cost, quality, human resource, communications, risk and procurement for the builder. The inclusion of the swimming pool at this stage will similarly also trigger the five project management processes as the project team work towards this addition.

b. Project scope management

Project scope management sets out in no uncertain terms what is included and excluded in the project. It defines the nature and size of the project and in so doing, determines what is necessary to manage the project to successful completion. An important outcome from project scope management is the creation of a work breakdown structure to define what must be done.

c. Project time management

Project time management requires the project manager to identify the tasks for the project, the time taken for these tasks to be completed as well as the sequencing of these tasks in a logical and productive manner. The end result of project time management is the creation of a project schedule with a time-line and milestones that must be controlled to meet the deadline set for project completion. In so doing, the various project activities must be identified and sequenced together with estimates of the durations needed for their completion. The project manager is then able to determine how long the entire project is likely to take and when different activities should be carried out.

d. Project cost management

Projects require resources. These resources include materials, machines, manpower, methods, management and money. Based on the activities needed for the project, the corresponding resources needed for these activities are then identified. The costs associated with the use of these resources are then estimated and summed to yield the project budget which must then be monitored and tracked to prevent over-spending. Project cost management therefore include planning, estimating, budgeting and controlling the costs of the resources used. It provides a clear picture of how much these resources cost and when the expenditure is likely to take place to avoid negative cash flow.

e. Project quality management

Project quality management deals with both services and products. Once the services and products for a project have been identified, plans must be put in place to ensure that the required level of quality will be met, together with the necessary steps taken to monitor that the outcomes meet the pre-determined criteria. Project quality management includes planning, assurance, control and measures for continuous improvements. Fundamentally, it defines and deals with how good the service or product quality should be.

f. Project human resource management

Projects are run by people and it is people that make projects happen. This is significantly an important area because getting the right people in a project sets the scene for what can be expected from the project outcome. Project human resource management identifies the manpower needed for the project, assessing who is best positioned to take on different roles, getting them on board the project, motivating them to deliver and evaluating their performance. Project staffing can be acquired from within the organization or out-sourced from external vendors on a contract basis. Organizational planning, establishing the organization structure and project team development are important issues for consideration in deciding who will do the work.

g. Project communications management

Just as cash flow is the lifeblood of the industry, it is communications that keep a project alive and moving. Without proper communications in place, no one can be sure of what is happening in the project at any point in time. Project communications management is about identifying, creating and disseminating the information needed by the stakeholders at the appropriate time. Such information does not simply appear out of nowhere. A conscious attempt must be made to ensure that the information needed is planned for, acquired and disseminated in an orderly manner using the correct mode. Through face-to-face meetings, email messages, formal reports, telephone conversations, minutes of meetings and letters, project communications management sets out the mode for correct and timely information to be delivered to meet the needs of the stakeholders.

h. Project risk management

Risks, if not managed properly, can lead to crisis and project failure. Project risk management, as a pre-emptive measure, therefore holistically identifies, measures, analyses and recommends appropriate mitigation steps to be taken in responding to project risks. In risk planning, it is necessary to strike a balance between the impact and the probability of a risk occurrence. This then sets the platform for the project manager to decide the planning rigor in which potential risks are to be identified, analyzed, monitored and controlled. At appropriate juncture, project risks can be mitigated, accepted, avoided or transferred to a third party. In so doing, the potential problems that may be encountered in a project can be identified early for their likely impact to be managed appropriately.

i. Project procurement management

Projects need resources for their completion. Such resources need to be procured and can include machines, materials, and manpower. The project manager needs to decide if such resources are to be procured from outside the organization or made within the organization. If these are to be procured from outside the organization, the vendors must first be identified and the proper organizational procedures such as the requirement for three quotations for open tenders should be followed. The project manager should also consider if the resources should be procured at the project level or centralized at the head office to benefit from better supplier discounts for bulk purchase. Project procurement management also deals with vendor negotiations, selection, and contract management. This knowledge area establishes what need to be procured for the project and from whom.

j. Project integration management

Projects are often peppered with changes or variations from start to completion. These occur for various reasons and can include unforeseeable budget cuts, introduction of new laws and regulations, exploiting the newest technologies, redesign to safeguard public safety and when the client changes his mind. All stakeholders are affected when changes occur. They must therefore be informed of such changes for them to take appropriate steps to respond to these changes. The project manager needs to ensure that changes are coordinated for disruptions to be kept to a minimum. Proper planning, execution and control are necessary to accommodate such changes. Fundamentally, when a change happens, the project manager needs to understand how this change may affect all the other knowledge areas, and what therefore needs to be coordinated and integrated.

1.6 Project Management Processes

a. The initiating process

The nine knowledge areas are underpinned by the five project management processes that include initiating, planning, executing, monitoring and controlling, as well as closing. These five project management processes normally take place sequentially even though there can be overlaps between processes when the situation warrants. In project procurement management, for example, this knowledge area must be initiated by the project manager, planned for, executed, monitored and controlled, and when procurement is completed, closed. At the project level, initiating means activating, triggering and doing whatever is needed for the project to commence with proper authority. At the activity level, for example in procurement management, the project manager must identify what needs to be procured for the project, and then initiate actions necessary to set this process in motion. Initiating sets the platform for planning to take place.

b. The planning process

Planning identifies all the activities and work that are required in a project for successful completion. What these activities and work are depend on the nature of the project. Planning is fundamental for success because it enables the development of appropriate policies, procedures and documents that cater to the activity and work so identified. For procurement management, for example, planning enables one to identify the materials needed for the project, the procedures for procuring these materials, the advance notice to be given to the supplier, and when these materials should be delivered to the site for installation.

c. The execution process

With planning firmly in place, execution can then take place uneventfully. Execution simply means carrying out the activity or putting labour and materials together to create the work needed for the project. The completion of execution leads to the completion of one milestone for the project. For example, if ready mixed concrete is needed for the construction of a high-rise building, the project manager needs to plan ahead to ensure that the correct quantities of concrete have been ordered and is available on that day, the concrete pump is in place to transport the concrete to the upper floor, and that the concreting team is present to place the concrete at the correct location. These tasks are then executed on the day of the concreting operations.

d. The monitoring and controlling process

Projects may veer out of their intended plans if not monitored and controlled closely. Hence, at both the planning and executing phase, monitoring and controlling are necessary to ensure that all parts of the projects fall in place to meet time, cost, safety and quality targets. While monitoring keeps track of progress, controlling helps to keep the project on track. For example, if concreting is to take place tomorrow, the project manager needs to monitor to ensure that all resources have been planned for and will be in place for concreting to take place at the appointed place and time tomorrow. Otherwise, corrective actions will need to be taken by the project manager to control deviations and mistakes. In the event that deviations and mistakes do occur, planning and execution will need to take place again until the work is completed satisfactorily, accepted and then formally closed.

e. The closing process

When an activity or work is completed, it needs to be accepted formally and the relevant stakeholders notified of the acceptance for the project to move on to another stage. Acceptance can take place after inspection of the work or product for satisfactory quality. Payments and proper documentation then take place and that part of the project is then considered to be officially finished and closed. Because many different activities and works can take place at any one time in a project, the closing process is also triggered throughout the entire project duration.

1.7 Areas Where Project Management can be Applied

a. Generic nature of project management body of knowledge

Projects can take place in any industries. These industries can include construction, transport, manufacturing, info-communication technology, hospitality, healthcare, finance, tourism, consulting, publishing research and development, and so on. A business that deals with production in the manufacturing sector may wish to construct new industrial facilities. This then amounts to a project for the builder, the building consultants and the manufacturer. A few years down the road, the manufacturer may wish to upgrade and retrofit the same industrial facilities. Once again, this is translated to a project for the builder, the building consultants and the manufacturer. Likewise, the concurrent design and installation of new production equipment by a specialist firm in the industrial facilities is also a project.

For this reason, project management principles are generic in nature and can be applied in projects across different businesses in different industries. In the construction industry, builders, architects, quantity surveyors and engineers undertake building projects. The project management principles that they apply are the same and cover the nine knowledge areas as well as the five project management processes.

b. Environmental influences

While there can be projects in different industries and businesses, the environment within which these projects take place can be vastly different. The physical site environment a project manager experiences in the construction industry is different from the office environment which another project manager experiences in the banking sector. The external physical environment may be linked to geographical, geological and ecological differences. Similarly, a project manager also faces a different set of social, professional, institutional and cultural environment from one industry to another. The social environment may account for different level of motivation demonstrated by different employees. The professional environment surfaces the different priorities of different professions such as between the accountant and the marketing director. The institutional environment highlights the different approaches to governance put in place by different local councils and countries.

As projects become increasingly global, more stakeholders with different nationalities are expected to work together. This can lead to cultural differences relating to values, beliefs and practices which a project manager must handle sensitively. When a project crosses national boundaries, issues relating to local and international politics come into play. The project manager must then pay attention to differences in religious beliefs, holidays, languages, foreign currencies and time zones. Hence, while project management principles are generic, the environment within which the project manager operates in can be vastly different. The environmental influence is an area that the project manager must be aware of for him to manage the project successfully.

c. Project management versus general management

The functions of the project manager and the general manager are different. Nevertheless, there are similarities between the two. Both need to have good interpersonal and social skills relating to motivation, communications and negotiation. At the same time, they must be problem-solvers who are able to demonstrate leadership to influence organizational outcomes. Where delineation is concerned, the project manager focuses on projects within the business while the general manager focuses on leading and guiding the business as a whole. The project manager applies the nine knowledge areas and is familiar with the five project management processes. In the course of his work, the project manager learns about and practises planning, organizing, coordinating and controlling. These are also the management skillsets of the general manager. However, while charting the overall strategic direction of the entire business, the general manager may not be familiar with the working details of specific projects that are related to scope, cost, time, quality, human resource, risk and procurement. He has to depend on the project manager for this purpose. The project manager may however rise up through the ranks to eventually become the general manager of the business. This is arguably the ideal situation where a project manager learns the ropes through completing live projects successfully before he progresses to the position of a general manager of the enterprise.

1.8   Beyond Project Management

a. Program management

It is normal for a very large project to consist of many smaller projects. These smaller projects work together to secure the successful completion of this very large project. Examples in the construction industry can include a super high-rise building or a mega size theme park. There are planning and design projects prior to constructing a super high-rise building. During physical construction, many projects take place at the same time on site. These projects include building the structural frame, installing the curtain wall, plumbing, electrical, fire protection, vertical transportation systems, interior decorations, as well as external landscaping works. At the same time, the project manager also needs to ensure that legal and regulatory requirements have been complied with through inspections. In large housing estates, these may be undertaken in phases. This is where program management comes into place where programs consist of multiple projects contributing towards a common goal. Hierarchy wise, these multiple projects or phases are managed by project managers who report to the program manager. The program manager ensures that all projects or phases in the program are well integrated and work together to complete on time and within budget.

b. Portfolio management

A portfolio is a collection of works of a similar nature. For example, an artist who specializes in landscape painting may have a portfolio of paintings related entirely to only landscapes. While a program is made up of several different projects or phases contributing to a common cause, a portfolio is a collection of projects of a similar nature. These projects of a similar nature may be guided by one line of specialty that the business focuses on. For example, a specialist supplier of a curtain wall product only deals with curtain wall and nothing else. The same curtain wall product can be installed in different residential, commercial and institutional projects. As far as the specialist supplier is concerned, it possesses a portfolio of projects that have installed its curtain wall product. Apart from similarity in the nature of work, a portfolio of projects can also be created based on a strategy identified by the business. For example, a portfolio of projects undertaken strategically in the capital city of a country under the direction of a regional director responsible for that geographical location.

c. Project management office

More often than not, businesses run more than one project at any one time. These projects may be parked under program management or portfolio management as appropriate. However, the head office of the business still needs to maintain oversight of all these projects. This is particularly important for large businesses operating in a large country with many of its projects that are dispersed geographically. The creation of a project management office in the head office can therefore serve to manage, organize and control all projects undertaken by the organization. There are however situations where it may be more strategic to have decentralized project management if it is in the interests of the organization to do so. For example, when the project concerned is located some 6 hour flight time from the home country, decentralization can expedite decision-making. Otherwise, a centralized project management office in the organization can help to facilitate better integration to minimize communication risks. Centralization can also help to pool resources for different projects to reduce costs. For example, in sharing a common pool of plant and equipment on a need basis for different projects. Furthermore, centralization can facilitate bulk purchase of materials for the organization to benefit from better discounts given by suppliers.

1.9   Revision Questions

1. Is project management new?
2. Who built the Great Wall of China and Egypt's ancient pyramids?
3. Who is the accidental project manager?
4. Can project managers be trained?
5. Do you think some people are born to be natural project managers?
6. Who make a better project manager? The architect, quantity surveyor or civil engineer?
7. What must we consider when project management take place in different sectors such as in manufacturing or in construction?
8. Can there be events within projects?
9. Is it true that operations can commence only with the completion of the entire project?
10. What do you understand by program management?
11. What do you understand by portfolio management?
12. What is the role of the project management office?

Chapter 2
Project Life Cycles, Stakeholders and Organizations

2.1 Introduction

a. Moving on from progressive elaboration and development

Projects do not just materialize from thin air. There must always be a purpose for initiating a project and for the project to be developed and refined through the process of progressive elaboration. For large and complex projects, an idea is first formed at the inception stage. The idea for a project is then progressively elaborated and developed to provide more details for implementation. For large and complex projects such as convention centres, theme parks and expressways, elaboration and development may take place in phases. These phases may be carried out sequentially if time is not of the essence or concurrently to fast track the completion of a project. The project life cycle is unlike nature where it is not possible to fast track the human life cycle or the butterfly life cycle. Fast tracking the butterfly life cycle as it evolves from the egg to the larva and pupa stage may cause defects in the butterfly. Throughout the project life cycle, people and organizations come into and leave the project. Suitable organization structures are therefore important to create appropriate behaviour to underpin project success.

b. Project phasing

Small projects do not require phasing and may be completed in one complete phase. However, large and complex projects have more uncertainties and should therefore be divided into phases to render the tasks more manageable. For example, a large township project may include residential buildings, shopping malls, schools, medical facilities, recreational facilities, offices, and roads. The developer may want to divide the township development into several phases, starting with the residential buildings as phase 1, shopping malls as phase 2, and schools as phase 3 and so on.

© Springer Nature Singapore Pte Ltd. 2018
L.S. Pheng, *Project Management for the Built Environment*,
Management in the Built Environment,
https://doi.org/10.1007/978-981-10-6992-5_2

Generally, one phase is completed before the next phase commences. However, when time is of the essence, phases may overlap one another to facilitate fast track completion. This means more activities are being carried out on site and thus heightening the probability of project risk occurrence. By dividing a large and complex project into phases, the project manager is then able to focus on managing one phase at a time to successful completion. Yet at the same time, the phases as a whole enable the project manager to view the project in its entirety. The uncertainties of a large and complex project can then be managed using the bite size approach. These uncertainties can be wide ranging and span over issues relating to scope, cost, time, quality, human resource, risk, communications, procurement and integration for the project.

c. Project life cycle

The life cycles of projects can vary depending on the nature of these projects. The life cycle is affected by the characteristics of the projects. Thus projects as diverse as the creation of a public park, cycling track, private hospital, and swimming complex have different life cycles. Typically, the project life cycle passes through inception, feasibility, construction, commissioning, hand over, maintenance and operations. Different projects therefore impose different demands on various stages of the life cycle. For example, the commissioning of a cycling track is not as exhaustive as the commissioning of a hospital with many sophisticated medical equipment and complex building services. Nevertheless, even though the project characteristics and life cycles are different, the project management activities that are carried out in both the cycling track and hospital projects remain the same. Both small and large projects need to deal with the various issues associated with the project management body of knowledge.

2.2   Project Feasibility

a. What project feasibility studies entail

It is worthwhile to reiterate that there must always be a purpose for initiating a project. It is not in the interests of the owner to commission a project without due consideration given to its feasibility. There must be a need or demand for such a project to render the venture viable. It should however be noted that some projects are profit-driven while others may be for a charitable cause. A project starts with the identification of a perceived need or market demand. A feasibility study is then carried out to assess if the need or demand can be met realistically through the project. In some cases, the feasibility study may uncover that such a need is already adequately met by a competing organization. The outcome of the feasibility study may then lead to the abortion of the project.

Alternatively, the idea may go back to the drawing board for a revamp before another feasibility study is conducted to once again assess the viability of the revised or totally new idea. For feasibility studies that deal with complex and wide-ranging issues, these may also be mounted as stand-alone projects. For example, the government may explore the possibility of privatizing some of its existing agencies. A consultant may be engaged to commission a study on the effects of the proposed privatization exercise. This study can on its own be a project in so far as the consultant is concerned.

b. Outcomes of feasibility studies

There can be many outcomes from feasibility studies. These can range from a simple yes or no outcome to back to the drawing board recommendations. From an organizational behavior point of view, some organizations may also commission feasibility studies to provide the desired outcomes to justify and support the decisions that they have already made. Such desired outcomes are then used as tools to pacify shareholders or convince the public at large of the viability of the proposed project. Upon completion of the feasibility studies, management should be provided with information on whether the idea or concept should now be adopted for implementation as a project. If the outcome is in the affirmative, further evaluation is still needed to determine if such a project is worthwhile to proceed in meeting organizational goals.

Such evaluation shall take into account the costs and benefits associated with the proposed project and how long this is likely to take for successful completion. To facilitate such an evaluation, further tests and experiments may be needed. A proto-type may need to be built for test-bedding and the associated costs to the end users noted. In some cases, the idea may be feasible but the proposed project may be halted at this stage because it does not fall within the core business of the organization. Similarly, the idea may be feasible to meet the needs identified but because of adverse environmental impact, the proposed project may also be halted at this stage. Yet in other cases, the adoption of the idea can only mean that the organization is diversifying its operations beyond its core capabilities. If the organization decides in the light of the feasibility study to proceed with the project, then it needs to assess how the project can be mapped to its needs and what further resources are needed to support this endeavor.

2.3   Project Life Cycles and Phases

a. Operations of the project life cycle

An analogy can be drawn between the project life cycle and the product life cycle in the marketing profession. In marketing, a product moves through the four phases of introduction, growth, maturity and decline. A product typically enters into the growth phase when it is still novel. However, over time when the novelty wears out, it reaches a plateau that signals maturity before going downhill at the decline stage. It is possible to reverse or delay this trend with constant upgrading of the product through the multiple product life cycle concept. The multiple product life cycle concept can also be extended when a matured product in one country is exported to another country where the product is still novel. Like the product life cycle with its four distinct phases, the project life cycle is typically also made up of phases. More phases are likely to be present in larger and complex projects. Large construction projects may also be divided into different zones according to locations. These zones are in turn subdivided into different phases. For large building projects, such phases may be divided distinctly into the foundation phase, superstructure phase, architectural phase, building services phase and external landscape phase.

Subdividing projects into phases allows the project manager to identify the nature and extent of work needed for the completion of each of these phases. This allows for the deliveries of correct materials, equipment, labour and other resources to be made to each phase. Phasing also allows the nine areas of the project management body of knowledge to be initiated at a manageable level for planning, executing, monitoring, controlling and closing. The completion of the foundation phase of a building project means that the deliverable expected relates to substructure works. It also facilitates decisions to be made that are associated with scope, cost, time, quality, human resource, risk, communications, procurement and integration for the foundation portion of the building project. Phasing also enables the project manager to determine which parts of the building project have the most risks and must therefore be given the appropriate dose of attention. The deep excavation and foundation phase of a building project typically carries more risks than the external landscape phase of the same building project.

b. Considerations for project phases

A project life cycle for a large project with a predetermined number of phases may take a longer time for completion if these phases are carried out sequentially. In most cases, this is because one phase must be completed before the next phase can commence. There is also the added constraint that some activities within a phase must, because of their nature, be carried out sequentially. For example, plastering cannot be carried out until the brick wall is completed. Similarly, painting cannot start until plastering and its drying out process have been completed. Where time is of the essence, phases may need to overlap to enable fast track completion for the project. Fast tracking may however lead to higher risks and costs with more resources being mobilized at the same time. Nevertheless, the resources needed tend to be less at the beginning of a project and pick up only as the project progressively and actively builds up steam. When the demand for resources increases for a project, correspondingly, costs also increase. The rate in which costs are expended in a project also varies with time and has often been described using the S-curve. Towards the end of a project, costs also tail off. Given the long durations of many projects, especially in the construction industry, the risks of project failure are very real indeed. Project failure may occur anywhere and for many reasons such as a devastating earthquake or financial crisis.

All things being equal, projects have less resources being expended or committed at the beginning and therefore face higher risks of being terminated at this stage. This is because the termination of a project at the early stage has less adverse impact on the stakeholders who may be more willing to let a project go in order to cut losses. Conversely, towards the tail end of the project life cycle, stakeholders have already invested much resources and are therefore interested to safeguard their interests. Hence, projects are more likely to fail at the beginning and less likely to fail towards the end of the project life cycle. In addition, stakeholders also realize that the project risks lessen as the project reaches the end of its life cycle. It is also at this stage that the stakeholders are not willing to jeopardize the risk level of the almost complete project by introducing changes or variations at this stage. Consequently, one can expect to see more changes and variations being introduced at the early stages of a project life cycle as these can be made more readily without correspondingly raising the risk level.

c. Example of product life cycle and phases in action

An example of how the project life cycle and its corresponding phases for a smart window is described. This starts with an identification of a need for a smart window that can block out daylight on a sunny day and let in more daylight during a cloudy day. At the same time, the smart window also doubles up as a solar panel to store clean and renewable energy that can be used to charge devices such as mobile phones. The need is then met with an appropriate invention that works. The invention is then patented. This then moves to commercialization. The first phase can be the proof-of-concept phase where it is established that it is feasible for such a smart window to be built. This then progresses to the next first-build phase where such a smart window is designed and put together to meet the specifications of the patent holder as well as the expectations of the stakeholders and management of this smart window project. Refinements and tweaking for improvements are then made at the third proto-type manufacturing phase.

This is where further tests are carried out to ensure that the attributes of the smart window have been translated to realistically achieve its declared capabilities to either admit or block daylight as well as store solar energy. In addition, further tests may be carried out in this phase to ensure that the smart window does not leak after full scale installation in a real building project. This may require the setting up of a full building façade in front of a large propeller fan with water dripping from pipes mounted at the top of the façade. The propeller fan and dripping water simulate driving wind and rain for a high rise building at the windward side. The smart window façade is then checked at the leeway side for possible leakage. If there is indeed leakage, it should be rectified through design refinements or sealants before the test is repeated until a set of perfect test-bedding results is achieved. This project has now reached the fourth phase of final build where the smart window and all the associated connecting parts are documented for flaws and the original proto-type adjusted to achieve the desired outcome. With this complete, the project moves to the last phase of operational transfer where full scale production of the refined smart window takes place. Each phase of the project life cycle has its own deliverables and in the smart window example, the five phases are carried out in sequence.

d. Advancing ahead with the project

The project manager is accountable not only for the timely completion of the entire project life cycle; he is also responsible for the successful completion of each and every phase in the project. Project performance may be evaluated at the complete project level or at the phase level. Fundamentally, the project shall be assessed for its performance in each of the phases using predetermined criteria for time, cost and quality targets. The project manager assesses if the project is on track for timely completion, within the budget and at the right quality standard. Projects are run by people who make this possible. Hence, an assessment shall also be made of the project team performance. As the project and the phases advance forward, the deliverables need to be shown to the owner to show proof of the completion of each phase.

While timely completion is important, the owner shall also be provided with the opportunities to verify or validate these deliverables to ensure compliance with the predetermined project scope of work. Verification can be facilitated through many ways. For example, in a building project, walk-throughs can be arranged for the owner or his representatives. Verification can also be provided by a third party audit conducted by the quantity surveyor representing the owner. The outcome of the verification exercise can lead to acceptance or feedback for rectification if defective works have been detected. Further verification exercises can be carried out upon the completion of the rectification works. Thereafter, proof of acceptance of the deliverables can be made in the form of approval documents or payment certificates made by the owner or his representatives.

e. Exiting one phase to another phase through stage gates

Upon successful completion of a phase, the project team exits from this phase and move on to the next phase of work. Exiting a phase may however culminate from successful completion or termination for failing to produce the predetermined deliverables. Exiting a phase will require the project team to pass through an imaginary doorway commonly known as the stage gate. Mentally, the stage gate is where the project team exits from one phase to another phase of the project. The use of the term stage gate simply means that the project team is only allowed to exit from one phase to another phase of the project when all the predetermined metrics have been met. These metrics can be related to quality, risks, scope and cost. For example, if there have been cost overruns that place the owner in a dire financial position, the project may be killed at this phase for lack of funding. Similarly, if the builder has been found to consistently deliver poor quality standards despite numerous warnings, the services of the builder may be terminated and the project team is then unable to exit from this phase to the next phase of the project.

Meeting the predetermined metrics is therefore a precondition for the project team to pass through the stage gate of a phase to proceed to the next phase of a project. Otherwise, revisions or rectifications are needed to make good or in the worst case scenario, the project may be terminated. A good example of a stage gate in building projects relates to the grant of the Temporary Occupation Permit from the building authorities. Upon completion of the building works, a Temporary Occupation Permit is needed for the building to be occupied. The builder needs to ensure that all fire, safety and regulatory requirements have been complied with. If these requirements have not been complied with, the permit will not be issued and the builder cannot then proceed to exit the stage gate of this phase to progress to the next phase of allowing the building to be occupied. In summary, when a phase is successfully completed, the project team exits the phase through the mental stage gate and progresses to the next phase.

The criteria or predetermined metrics for phase exit depend on the type and nature of projects. Different performance metrics may be used for different purposes covering issues relating to quality, security, public health, environment, legislation, and proper workings of the plant and equipment installed. Audits, inspections, walk-throughs may also be conducted to ensure that the owner takes possession of what was earlier agreed upon contractually. Once this assurance is secured, the owner then signs an acceptance notification to allow the project team to move from this phase to another phase of the project. This also marks the successful completion of the phase.

2.4   People like You and Me

a. People as stakeholders

People are the stakeholders who drive projects. As stakeholders, they may come from a variety of organizations that are actively involved in the project and in turn, can be affected by the project outcome. Stakeholders are fundamentally people who have vested interests in the project and may act singularly as an individual or collectively as a group. They may be internal or external parties to a project and can include customers, communities or interest groups. While most stakeholders generally support a project, there may be some stakeholders who object to the project. In the latter case, for example, a community may object to a neighboring project if there are potentially adverse environmental impacts arising from the project.

It is therefore crucial for the project manager to identify and recognize who the various stakeholders are to understand their needs and concerns as early as possible at the project inception stage. This is necessary to manage stakeholder expectations which may differ from individuals and groups. Early management of stakeholder expectations is necessary to secure their buy-in for the project. However, not all the demands or requests of stakeholders are reasonable or workable because of budget, time and technical constraints. In such a situation, the project manager needs to discuss and negotiate with the stakeholders concerned to arrive at a consensus and to promptly put the agreements in writing to better manage stakeholder expectations.

b. Stakeholders in construction projects

Projects in the construction industry can range from simple renovation jobs to mega shopping malls. Correspondingly, the number of stakeholders can vary from project to project. Construction projects are also short term in nature. While a renovation job can be completed in one month, the construction of a mega shopping mall can take up to three years. Regardless of its size, stakeholders in a construction project do not work together in the long run. They come together for a project and when the project is completed, disband and return to their own organizations or move on to another project. Large construction projects are generally complex in nature and involve many stakeholders. They can include the project manager, property developer, main contractor, suppliers and subcontractors, building authorities, real estate agency, tenants and property purchasers.

Within each organization, there are also many internal stakeholders who are involved with projects. For example, the client in a government ministry that manages school projects may also set up a project management office comprising project managers, who look after different school projects. Apart from direct stakeholders, there may also be indirect stakeholders who can both positively or negatively influence a construction project. The indirect stakeholders can be members of the local nature society or charity groups. Large construction projects are generally more complex than projects in other industries because of their nature and magnitude involving a lot more stakeholders. Project managers who have successfully completed large construction projects can therefore transfer their skillsets to other industries readily.

c. What can stakeholders do?

There are many ways in which stakeholders can influence a project. Some of these influences can drive a project while other influences may hinder the progress of a project. At the community level, stakeholders may leverage on political power to influence the project outcomes. For example, a political leader may promise to build more facilities for a community if the people vote for him. Stakeholders can also demand changes to either add to or remove features from the project scope. A community or environmental interests group may ask for the construction of a road to be diverted to avoid too much disturbance to the surrounding natural habitat.

The parent support group of a school may request for additional features in a school upgrading project to include a swimming pool or a rock-climbing wall. While stakeholder contributions are generally positive in nature, stakeholders can also inflict negative influence through grapevine gossips and unfounded rumours. In extreme cases, disgruntled stakeholders may actually sabotage a project through physical acts of vandalism. For example, stakeholders living near a proposed rare earth mining project may pass rumours about the pollution created and potentially harmful effects of radiation on the neighbourhood if the project is to proceed. For health reasons, the stakeholders then mount pressure on the government not to approve the project.

2.5 Organization Structures

a. How are projects organized?

Projects can be organized in many ways. There is no one best way to organize a project. The approach taken to organize a project depends on the circumstances of each particular case. However, in all cases, projects are parts of larger entities set up to provide a unique product such as a new building or a unique service such as a new call centre. These larger entities can include government ministries, companies, charitable organizations and neighbourhood communities. They initiate the projects and therefore have a direct influence on these projects. For example, the public housing agency is set up in a country to provide affordable mass housing for the population. The agency can decide where the housing estates are to be located, their respective architectural designs and who the builders are. The various public housing projects are therefore initiated by the agency who decides how these projects are to be organized and managed.

There are three levels of organization emanating from the larger entities to their projects. These three levels are the Executive level, the Functional level and the Operations level. In the example of the public housing agency, the Executive level provides the vision, mission, policies and strategies to achieve the wider goal of providing affordable housing to the population. The Executive level decision-makers can be government ministers who provide answers to questions on why the population needs to be adequately housed to ensure political stability which is a precursor to economic growth. Issues relating to quality of life, high-rise or low-rise building options and floor areas of apartments for different family size are debated at the Executive level. Once the vision and mission are established, these are operationalized into policies and strategies for actions to be taken.

These then move to the next Functional level where a clear understanding of the project purposes is set out, and operational tactics are conceived and put in place to accomplish the public housing objectives. Functional departments and corresponding business processes may then be set up for this purpose. These departments can include the Construction Department, Human Resource Department, Legal Department, Finance Department and Contracts Departments all with their respective functional roles and responsibilities. These departments may be established in-house within the agency or out-sourced as appropriate. For example, to ensure greater varieties in building designs, these may be out-sourced by the government agency to private sector architecture firms or design-and-build construction firms. Projects are then carried out at the Operations level which involves the day-to-day routine running of the projects from inception to completion.

The construction firm tasked to build a specific public housing project runs the operations that may include decisions on how the building is to be constructed. For example, choices have to be made between different construction technologies relating to cast in situ or precast concrete and various formwork systems. The Executive, Functional and Operations levels do not exist only within the client organization which in this case is the public housing agency. The same three levels also exist within the construction firm which similarly needs to deal with questions relating to the why, what and how for the projects that the firm undertakes.

b. How organization structures influence authority of the project manager

Projects may be organized using either a Project structure or a Functional structure. In between the Project-Functional continuum, projects may also be organized using a Matrix structure. However, the Project structure and the Functional structure may co-exist within an organization. For example, the head office of a construction firm may be organized using the Functional structure of which there is a Construction Department. The Construction Department may in turn be organized based on the projects that it currently undertakes. In a Functional structure, project managers report directly to the functional manager and to avoid confusion, are often called project coordinators or project team leaders. Similarly, for large organizations that run mega projects, a Functional structure may also be present in one of these mega projects. For example, during the construction of a mega shopping mall, the builder may organize the project team according to functions such as safety, quality, contracts and human resource who report to the project manager.

Hence, the choice of either the Functional structure or the Project structure can determine whether the functional manager or the project manager can exercise more power and authority in the project. In the Matrix structure, both the functional manager and the project manager may share equal power and authority if the matrix structure is balanced. Tilting to either a strong or weak matrix structure may mean the project manager has respectively more or less authority relative to the functional manager. Apart from power and authority, the Matrix structure also suffers from lack of clarity for employees who need to report to both the functional manager and project manager at the same time. This arises if there are conflicting instructions given by the functional manager and the project manager to the same employee. Such a situation may potentially give rise to office politics caused by poor communications and misunderstandings.

c. How organization structures communicate

Communications is the highway of projects. There must be a continuous flow of timely and relevant information for project team members and stakeholders to coordinate their efforts. Information has often been described as power. How the project is organized can therefore determine how the information flows. In essence, a Functional structure will route information flows through the functional managers. Likewise, a Project structure will route information flows through the project manager. However, for large organizations, such information flows may not be so straightforward. This is because information may be lost or distorted at the interfaces between functional departments or projects. For example, there may be a Construction Department among the several functional departments in the head office of a large construction firm.

There can be more than one project team reporting to the functional manager of the Construction Department which also oversees the pooling of the firm's plant and equipment resources. The functional manager assigns the plant and equipment to the various building projects that the firm undertakes. Clear communication, coordination and information flow between the functional manager and the various project managers are therefore crucial for the firm's success. This is where information flows between the functional manager and the project managers take place. In assigning specific plant and equipment to a particular building site, the Safety Department in the firm needs to be informed to check for safe installation and use on the site. Hence, there is also information flow between the Construction Department and the Safety Department within the firm.

d. Hybrid structures

Projects are generally organized using the functional, project or matrix structures. Nevertheless, there can be circumstances where special projects are organized using hybrid structures which are basically combinations of different structures. These may be projects that are one-off and of much strategic significance to the organization. A hybrid project team is then formed for this purpose. Members of the hybrid project team can be drawn from various functional departments that already exist within the organization. These functional departments can include the Construction Department, Safety Department, Contracts Department, Legal Department and Finance Department. The talents from these functional departments are then temporarily seconded to the project team that is headed by a project manager appointed for this purpose. The hybrid team members work together and are then disbanded to return to their own functional departments upon completion of the one-off project.

Team members are generally selected for their expertise that they can bring on board the project. In some cases, team members are assigned to the hybrid project team while at the same time, still retaining their works in their respective functional departments. For example, the chief executive officer of a construction firm may wish to implement high-priority work improvement practices across the various functional departments in the business. However, seeing this work improvement project that cuts across different functional departments and is of a temporary nature, it may not be appropriate to set up another permanent functional department just for this purpose. The chief executive officer may then choose to adopt a hybrid structure by appointing the current quality manager to double up as the project manager for the work improvement project. Employees from the various functional departments are then assigned to the work improvement project while still retaining the existing appointments in their respective departments.

This hybrid structure can help to create synergy for organizational learning by getting different functional departments to learn from each other. These employees bring to the work improvement project their expertise on how best the exercise can be implemented for the business. They also bring the lessons learned back to share with their colleagues in their respective functional departments. The work improvement project may last for a few months and is then disbanded once its aim and objectives have been achieved. Hence, there are different ways to get projects up and running. There are also different ways for projects to be organized.

2.6   Revision Questions

1. Who can initiate a project?
2. In what ways can projects be identified for initiation?
3. Can a stage gate be physical in projects?
4. Can phases in projects be predefined?
5. Are all phases in the construction industry typically similar?
6. What do you find in phases?
7. Are there differences between Zones and Phases in construction projects?
8. Is it possible to have one project manager for each phase of a large construction project?
9. Can sales of a large private condominium project be in phases?
10. In what ways can the marketing product life cycle be renewed?
11. Can project life cycle be renewed?
12. Who drive the project life cycle?
13. Why are different stakeholders involved with different stages of a construction project life cycle?
14. Can project life cycles be replicated in the construction industry?
15. What are some of the reasons for terminating projects?
16. Is it true that organization behavior is expected to be different in different organization structures?
17. Is it possible to have hybrid organization structures in a company?
18. Are Functional structure and Project-based structure mutually exclusive?
19. When do we use program management?
20. When do we use portfolio management?
21. When do we set up a project management office?
22. Can project management, program management, portfolio management and project management office co-exist in an organization at the same time?
23. Is there a best structure to accommodate increasing multi-nationalism in a construction project?

Chapter 3
Project Management Processes

3.1 Introduction

a. Processes in projects

All projects are made up of processes that fall within the nine areas of the project management body of knowledge. To reiterate, these nine areas relate to integration, scope, time, cost, quality, human resource, communications, risk and procurement. There are five processes that kick-start each of these nine areas. These five processes are initiating, planning, execution, monitoring and controlling, and closing. The five project management processes are related in some ways to general management principles that cover planning, organizing, coordinating and controlling. However, in the project management domain, the five processes are adapted for each of the nine knowledge areas as well as for different industries and businesses.

Hence, the five processes and nine knowledge areas are equally relevant for the project manager in the construction, manufacturing, transportation, hospitality, banking and healthcare industries. Take for example, project quality management in the construction industry. Quality management does not simply materialize out of thin air. Someone must trigger the processes for quality management to take place within construction projects. The project manager must initiate the process, plans for the quality management exercise, ensures its proper execution, monitors and controls compliance along the way, and proceeds to closing when all quality targets have been met. It is generally acknowledged that the planning process is the most intense process in project management.

b. Where the processes featured

The five processes take place in projects that are both small and large in magnitude. More of such processes are expected to take place in mega projects that involve many project team members and stakeholders that cut across multiple phases in such projects. For example, a mega integrated resort project may include the construction of several hotels, different theme parks and a casino. This mega resort

© Springer Nature Singapore Pte Ltd. 2018
L.S. Pheng, *Project Management for the Built Environment*,
Management in the Built Environment,
https://doi.org/10.1007/978-981-10-6992-5_3

project can be divided into phases for implementation. Phase 1 can be for the hotels, Phase 2 can be for the theme parks, and Phase 3 can be for the casino. Each phase can in turn comprise of different projects. For example, there can be Hotel A, Hotel B and Hotel C in Phase 1. Each hotel project in turn has its own work breakdown structure and the various operations dealing with different building elements and trades.

There are similar issues relating to integration, scope, time, cost, quality, human resource, communications, risk and procurement for all building elements and trades associated with each of these projects in the different phases. The five project management processes are therefore featured across all phases, projects, work breakdown structure, building elements and trades in the mega resort project. This helps to ensure that the targets set for the nine project management knowledge areas are met for the mega resort project. In totality, the processes for a large building project can be complex and involve many different areas of work.

c. What the processes involve

Projects are made up of processes. Processes in turn are made up of a series of actions and activities that sets out to meet the common goal of the parent project. Projects are therefore completed through these processes. The processes move the projects and their phases forward through initiating, planning, executing, monitoring and controlling, and closing. Take project scope management for example. The scope is identified progressively through a work breakdown structure. The work breakdown structure for a building project may be divided into the substructure work, the structural work, the architectural work and the building services work. The components that make up each of these works must then be identified. For example, the components for substructure work can include excavation, piling and the construction of the basement. Similarly, the actions and activities needed for the completion of each of these components must be identified. For example, activities for the construction of the basement can include excavation, formwork erection, waterproofing, installing the reinforcement bars, concreting and backfilling.

For a typically large building project, the magnitude of the scope, work breakdown structure, components, actions and activities can be magnified many times involving many different stakeholders both internal and external to the project organization. It is also often the case that the project management processes go through an iteration process as more information and firm decisions are progressively made available. The five project management processes generally evolve sequentially. However, the planning, executing as well as monitoring and controlling processes may iterate for various reasons. These reasons can include the availability of new information, new regulations, changes requested by the client or defective works. In such a situation, the project manager takes the task back to the drawing board to plan for the next course of action. The decisions made for planning can in turn influence the execution as well as monitoring and controlling processes of the project.

3.2 Process Groups

a. How the process groups work

The five processes of initiating, planning, executing, monitoring and controlling as well as closing are not individual and solo activities. These processes operate in groups as applied to various project management knowledge areas such as time, cost and quality. As a collection of activities, these process groups serve to implement and control the project life cycle as it moves from one phase to another. Inevitably, changes or decisions made in one process can affect another process within the group. Similarly, changes in the process groups for one project management knowledge area can likewise affect another knowledge area. For example, if the painting for a plastered wall is found to be defective in a building project through the monitoring and control process, the paint will need to be scraped and rework is thus necessary.

The repainting activity will need to re-iterate through the planning, execution as well as monitoring and control processes. The defective painting work also affects the project management knowledge areas of human resource, procurement, communications, time, cost and quality for the building project. Integration is therefore crucial in such a situation to ensure that all stakeholders have been informed of the delay caused by the defective paintwork. In the construction industry, the output of one process group for a certain trade also serves as an input to another succeeding trade. For example, painting can commence only after plastering has been completed. In this case, the process groups may work sequentially if there is no overlap.

b. Sequential or overlapping processes

In projects where the various work breakdown structures are likely to be similar, their corresponding processes can be replicated. For example, in the construction of a simple ten-storey office building with similar floor layouts, the activities of plastering and painting are more often than not, likely to be repeated in each floor. In such projects, the process groups associated with plastering and painting can be repeated for the activities associated with planning, execution as well as monitoring and controlling. It should also be recognized that such process groups have implications for the knowledge areas relating to cost, time, quality, human resource and procurement for these plastering and painting trades. Similarly, there are also projects where time is of the essence and the client may request for fast-tracking to shorten the overall project duration.

In this case, the work breakdown structure, components, actions and activities for the project remain the same except that more resources may be poured into quicken the project pace. This is where there are overlaps in the various process groups for different building elements or trades. Apart from the beginning of the initiating process and the tail-end of the closing process, there can be overlaps between some parts of the initiating and closing processes with planning, execution as well as monitoring and controlling. Overlapping of processes has implications for integration, time, cost, quality, human resource, communications and procurement. If the overlap is massive, project risks may heighten. Project management processes may therefore be sequential, overlapping, repetitive or iterative in nature.

3.3 The Initiating Process Group

a. Identifying the needs

The initiating process can take place firstly at the project level and secondly at the level of the various project management knowledge areas. Initiating at the project level must take place with the project need identified before triggering the various project management knowledge areas. A project does not simply appear out of nowhere. Someone must have identified the need for the project which can be linked to the development of a new product or offering of a new service. The identification of a project can be linked to the marketplace where a new product or service is created to meet market needs. The new project can be a response to fulfilling these needs. The new project can take advantage of an opportunity or offer a solution to a problem. However, not all new projects are profit-driven. Some new projects may be started by charitable organizations to help the poor and needy. Similarly, some non-profit projects may also be initiated by private sector businesses to fulfil their corporate social responsibilities.

Some new projects may not be based entirely on new discoveries. Instead, these new projects may be incremental inventions and based on marrying two or more existing products or services. There are many avenues to take advantage of an opportunity to meet market needs and demands. These may include initiating projects that aim to reduce administrative costs, enhance revenues, and eliminate wastes to improve productivity or simply to meet a societal need. Many opportunities abound for those who are enterprising enough to identify these needs. For example, a property developer may recognize the greying population of a community and proceeds to build more homes that are specially designed for the elderly. In another case, a travel agency may recognize the abundant time that the elderly folks have and proceeds to offer more overseas travel programs that cater specially for them. These are projects for both the property developer and the travel agency based on needs that they have identified from the marketplace.

b. Is the project viable?

Identifying the needs and the opportunities is only the first infant step towards turning the project into a reality. While needs and opportunities can be identified, the proposed project in response may well turn out to be unworkable for various reasons. These reasons can include excessive costs, unattractive profits, overestimated demand, or the project is simply ahead of its time. It is therefore important to conduct a feasibility study to ensure that there is actually a problem whose needs the proposed project can help to meet. The feasibility study can be conducted in various ways. These can include market surveys, interviews with potential customers, simulation exercises and laboratory experiments. Current regulations that may potentially impede the proposed project should also be studied at this stage.

The procedures and results of the feasibility study should be documented with the real problems and needs clearly identified. It should also document the opportunities identified with concrete proposals on how these needs may be met realistically with the current technologies available. A comparison should also be made between the costs of creating the solution to the rewards that the project may potentially reap. It should be recognized that some aspects of the feasibility study cannot be quantified such as those relating to how the proposed project is likely to impact the community at large from a societal perspective. In such a situation, the comparison should be undertaken honestly to avoid misleading recommendations. The outcomes of the feasibility study may lead to tweaking of the proposed project, postponement or abandonment.

c. Creating the product or service description

Following the completion of the feasibility study, a decision may be made to proceed with further progressive elaboration and development of the proposed project. Creating a description of the new product or service is a good exercise for stakeholders to conceptualize what they have in mind and expect from the project outcomes. If the outcome is a product, then that product should be described. An example of a brief product description in the construction industry can be the proposed erection of a three-storey detached house with swimming pool and car porch. A construction consultancy firm can create a service description to offer an all-inclusive building information modelling package for potential clients. However, not all product or service description can be created that readily. In some cases, the description relates to a desired future state for a service that can only be experienced by the customer at a later stage. A travel agency may provide a write up on a proposed ten-day tour package to Europe. The tour itinerary created by the travel agency describes to the customer what can be expected of the tour package on a day-to-day basis in the near future.

d. Creating a preliminary scope statement for the product or service

The creation of the product or service description sets the platform for the project team to further refine and delineate what is included and excluded in the product or service offering. This refinement and delineation exercise is carried out using a preliminary scope statement. Information and details may be unavailable at this early stage of the project. Hence, the scope statement can only at best be preliminary in nature. With time, when more concrete information and details are made available to the project team, the preliminary scope statement can be further expanded to form the project scope statement. The preliminary scope statement is useful for the project team to start thinking about what needs to be done to achieve the desired project outcome. However, it should be noted that the preliminary scope statement should not be interpreted to merely mean a statement or a paragraph.

Depending on the nature of the project, the preliminary scope statement may consist of a small bundle of documents that include floor plans, architectural models, specifications and a fairly detailed description of the proposed project. The preliminary scope statement defines what the project is to accomplish and the expectations of the various project stakeholders. The small bundle of documents should at the very least identify the project vision and mission, product specifications, deliverables and their acceptance criteria, boundaries of the project, the people involved and how future changes are to be reviewed and approved. There are no set rules on what should be included in the preliminary scope statement. The rule of thumb should be that as much information that is already made available should be provided as far as possible. The earlier the project team is clear about the project scope, even if it is only preliminary, the better it is for the project.

3.4 The Planning Process Group

a. Introduction to planning

Planning is by far the most intense and time-consuming part of project management. It is generally acknowledged that if one fails to plan, one plans to fail. The process of planning shows that the project team has spent time and efforts to think through as far as possible, all pertinent aspects of a project and the corresponding issues that may arise. Planning is rigorous because it combs through the nine areas of the project management body of knowledge that encompass integration, scope, time, cost, quality, human resource, communications, risk and procurement. The planning process is generally iterative in nature. It is always good to plan early. However, early planning suffers from lack of information. As time progresses, more information is available and the project team therefore needs to iteratively update the project plans.

Information can come in the form of alternative building technologies that may be adopted for a new construction project as well as the costs, quality, risk and time associated with the use of a specific technology option chosen for the project. The planning process should as far as possible include all the stakeholders to obtain early buy-in of the various project deliverables. Apart from the client and project team, other stakeholders include the suppliers, subcontractors and government agencies. The inclusion of stakeholders in early planning allows the project team to understand their requirements to better manage their expectations. The project team should also recognize that not all customers are external customers; some customers can also be their internal customers working within the same organization. The voice of the customer should therefore be heard for planning to take place smoothly and correctly. Methodologies such as the quality function deployment technique can be used to facilitate the planning process. In mega projects, commercially available soft-wares can be used to collate and keep track of the thousands of tasks and activities involved in the project.

b. Developing the scope statement

A preliminary project scope statement was earlier created in the initiating process. This preliminary project scope statement is then progressively elaborated and developed further in the present planning process. The project team should possess the necessary experience to understand what is needed to complete the project. For example, if the project relates to the construction of a mega integrated resort project, project team members are expected to possess the necessary knowledge on what such a project entails and what its various component parts are to facilitate the planning process. In exceptional cases where the project team members do not possess the necessary expertise because new cutting-edge technologies are involved, then they should obtain such expertise from the relevant external consultants.

To facilitate planning and developing the scope statement, project team members should be competent not only in project management skills, but should also be familiar with practices and technologies used in the chosen discipline. This is because while project management skillsets are transferable, the domain knowledge in a chosen discipline cannot generally cut across industries. For example, it would be difficult for a project manager in the information technology sector to develop from scratch the scope statement for a new building project in the construction industry. Planning for and developing the scope statement require project management skillsets as well as relevant domain knowledge. With such background competence, the project team members can then develop the scope statement showing how the project scope is to be defined and what the project includes as well as excludes.

Once the scope has been defined, changes, while allowed, should be controlled. Without appropriate change control, it is difficult to freeze the project boundaries to categorically describe the work necessary to meet the project objectives. Uncontrolled changes can lead to delays and disputes. Hence, the project team should set out as early as possible to determine the parameters within which changes will be allowed or disallowed. Nevertheless, in describing a desired future state as part of the description for a service offering, disclaimers have often been used by the service provider to the effect that changes will be made without further notice. Such disclaimers have been used, for example, by travel agencies where tour itineraries need to be changed on the spot because of poor weather conditions.

c. Developing the work breakdown structure

Following the creation of a clear scope statement, the project team can now proceed to develop a work breakdown structure for the project. This is an exercise where the interaction between the project management skillsets and the domain knowledge is most significant. The work breakdown structure provides a systematic and organized approach to understanding what the project deliverables are. These deliverables are generally organized in a format typically recognized and practiced in the industry concerned. For example, in the construction industry, a typical work breakdown structure for a simple house dwelling project can consist of substructure works, superstructure works, architecture works, building services as well as furniture and fittings.

Hence, where appropriate, the project team need not reinvent the wheel when such a typical work breakdown structure format already exists in the industry that can be readily adopted and tweaked for use. When completed, the work breakdown structure provides a list of the components, actions and activities required for successful project completion. In turn, the list of components, actions and activities provides the inputs for the project team to plan for integration, time, cost, quality, human resource, communications, risk and procurement. Hence, the more detailed the work breakdown structure is, the less likely it is for the project team to make errors in these areas.

d. Developing the network diagram

Understanding what the components, actions and activities are for the project is the first step for the project team to determine the time needed for their completion. The work breakdown structure provides the list of activities required for the project. With the activities identified, the project team needs to plan for when and how these activities are to be carried out. The methods and approaches used to complete these activities must however be first determined. This is because activities can be completed using different methods. For example, a building project can use either cast in situ concrete or precast concrete for constructing the structural system. The choice can be affected by cost and quality considerations. In some cases, there is no choice at all because a particular method of construction is already mandated by regulations.

Determining the methods used to complete the activities allow the project team to estimate the time needed for their completion. Time estimates are therefore established for all activities to further determine how long it takes to complete the entire project. These activities are then sequenced in a logical manner to facilitate smooth work progress. This is often sequenced in the manner in which work is expected to be completed. For example, superstructure works should generally commence only after the completion of the substructure works. Nevertheless, there are also options available such as the top-down approach that allows the construction of the substructure works and the superstructure works to run concurrently instead of sequentially. Having established the list of activities and sequencing their logical work flow, the project team can proceed to develop the network diagram for the project.

The network diagram shows all the activities in the order in which these will be carried out either sequentially or concurrently. Apart from showing the relationships between different activities, the network diagram also identifies the overall project duration as well as the critical and non-critical activities of the project. Critical activities play a significant role in determining the overall project duration. Hence, critical activities must not be delayed in order not to delay timely completion of the project. On the other hand, non-critical activities can be delayed to some extent without affecting the overall project duration. This is because of the availability of float time in the non-critical activities which allows the project team to somewhat delay the start of these activities. Nevertheless, non-critical activities should not be overlooked to such an extent that their delayed completion can in turn lead to them becoming critical and thus affecting timely project completion.

In cases where the client wishes to complete the project in as short a time as possible, the project may take on a fast-track mode with more activities taking place at the same time. Having a deadline set by the client, means that the project duration needs to be compressed. Having too many activities running at the same time concurrently may however also heighten project risks. It is not unexpected that many activities, running into the thousands, are needed for large complex projects. The large number of activities can be so massive that it is impossible for the human mind to comprehend and monitor. In such a situation, commercially available software programs should be adopted for developing the network diagram.

e. Completing the cost estimates

Having established the work breakdown structure, chosen the most suitable methods to carry out all the activities identified and created the network diagram, the project team can then proceed to determine the cost estimates. This can take place at the level of the activities and the estimates of all activities are then summed to determine the overall project cost estimate. Estimates for activities can be built up in-house from first principles using unit rates for the resources needed for these activities. Estimates may also be obtained for activities that are outsourced to external vendors and subcontractors. It should be recognized that cost estimates do not just focus on the direct costs of the resources needed for completion of the activities.

Some activities carry attendant risks and the costs associated with the risk mitigation measures such as insurance premiums must therefore be factored into the estimates. In some cases where it is not clear if additional costs are likely to be expended, the project team may set aside a contingency or provisional sum for this purpose. There are many ways in which cost estimates can be calculated. Firstly, this can take the form of top-down estimates with the client setting a ceiling which the project budget is not to exceed. For example, the client may specify that the total amount he is willing to spend to refurbish his detached house should not exceed a certain amount. With this ceiling in mind, the project team then prepares the cost estimates for all activities and at the same time, balancing quality standards with costs.

Secondly, in the case of the construction industry, cost estimates can be prepared by the contractor's estimator or the consultant quantity surveyor. This generally takes the bottom-up approach with the costs associated with each activity estimated based on the quantities of and the unit rates for the resources used. Some examples of unit rates can include $ per cubic metre for ready-mixed concrete, $ per square metre for ceramic floor tiles and $ per metre run for copper piping. The bottom-up approach is fundamentally more accurate in producing the cost estimates. In the construction industry, for large projects, it is often the practice for the client to engage a consultant quantity surveyor to provide cost advice and to serve as an independent check as to whether the bids submitted by the builders are fair and reasonable.

Thirdly, cost estimates can also be determined informally through business associates and colleagues within the same organization. It is quite often the case that the nature of the project is not new to the organization. For example, a construction firm may choose to specialize in school building projects. The firm has already completed several school building projects in recent years. Hence, when the firm bids for another school building project, there is already a store-house of cost data and estimates relating to the construction of school building projects that the firm can rely on. In some cases, the client may approach a consultant quantity surveyor to informally enquire about the estimates for a new hotel project. Based on similar hotel projects completed in the past, the consultant quantity surveyor may informally provide the cost estimates for the new hotel project based on a $ per hotel room basis.

f. Developing the project schedule

There is a close relationship between the creation of the network diagram and developing the project schedule. The network diagram shows the relationships between the various activities that are carried out either sequentially or concurrently to complete the project. Upon completion of the network diagram, the project schedule can then be established. The project schedule shows how long it takes for the project to be completed based on the work breakdown structure. The project schedule can either be established through the network diagram or stipulated by the client. If the time-line is imposed by the client, the project team needs to work backward to determine how best the various activities are to be carried out to meet the stipulated deadline. Where necessary, some activities need to be crashed or compressed to shorten the project duration. However, if the client does not impose a strict deadline for project completion, then a fair and reasonable project schedule can be established objectively based on the network diagram developed earlier based on the work breakdown structure.

The critical path in the network diagram is the series of activities in the project that cannot be delayed without delaying the project end date. There can be more than one critical path in a project and care must be exercised to ensure that all activities on the critical paths are completed on time. This is because there is no float or slack time for these activities and any delays in their completion will inevitably lead to delay in project completion. However, while activities on the non-critical paths have some float or slack time, these activities should not be delayed unduly until they in turn become critical. Due to various unforeseen circumstances such as inclement weather, it is quite common for critical paths to shift from one series of activities to another. An earlier non-critical path may therefore be rendered critical. This may cause delay to the project schedule and in some cases, lead to the imposition of liquidated damages for delays in completion and hand-over.

g. Planning for project financing

Completing the cost estimates and project schedule enables the project manager to understand how the monies are to be expended for the project. The project manager should try to balance expenditure with income to ensure that as far as possible, there is positive cash flow for the project. Project budgeting therefore deals with the expected costs of the project in tandem with the cash flow projections. The process allows the project manager to understand how much and when monies will be spent throughout the entire project duration. In the construction industry, progress payments for the builder are often made on a monthly basis. In between these monthly progress payments, the project team needs to fully grasp the cash outflows and inflows to ensure that the project accounts do not go into the red.

If there is indeed negative cash flow, the builder needs to either top up the difference using his own funds or if he has good credit rating, borrow from the banks as a last resort. However, borrowing from the banks may further erode his financial standing because of the interests payable to the banks. In some countries, the builder may resort to delaying payments to the subcontractors or suppliers to ease his own cash flow. However, this is no longer possible in countries where there are legislative provisions made for timely payments to the subcontractors and suppliers under the Security of Payment Act. In project budgeting, the cash flow projections will therefore allow expenses and income to be planned. In the construction industry, this frequently follows a S-curve, with cumulative expenses being lesser in quantum early in the project life cycle. These expenses then take on a steeper curve as the project moves into full swing, and then tampers off towards the tail end of the project.

The client should also be aware of the bidding strategies that the builder may adopt when formulating the bid price and be watchful of front-loading of unit rates made by the builder to ease his cash flow. In the construction industry, it is also common for the client to retain ten percent of monthly progress payments to ensure that defective works, if any, will be made good by the builder. Half of the retention monies will be returned to the builder at the end of the contract. The remaining half is retained for the one-year duration of the defects liability period to ensure that the builder makes good any defective works that are uncovered after the completed building project has been handed over to the client. These contractual provisions do play a part in affecting project financing and should therefore be factored in by the project team for cash flow projections.

h. Developing the quality management plan

The project scope statement and the work breakdown structure identify the components, actions and activities needed to complete project. Some of the works involved are tangible products such as doors and claddings. Other works involved are intangible services such as the workmanship in which the ceramic floor tiles are laid or the service level of the real estate marketing agent. The development of the quality management plan is therefore an important first step to ensure that the desired quality standards have been delivered at all stages of the project. Delivery of the expected quality standards is often a condition precedent for progress payments to be made by the client to the builder.

The quality management plan for the project shows how the quality issues for the project are linked to the organizational quality policies. These policies may be related to existing ISO 9000 quality management systems, total quality management principles or six sigma metrics. The project team should recognize that it is cheaper to deliver the expected quality standard right the first time all the time. Failure to deliver the expected quality standard will tarnish the reputation of the organization. In addition, there will be additional costs incurred as well as time delays in correcting defective workmanship quality or products. The project team can refer to existing standards or codes of practice in the relevant industry to benchmark the expected quality standards. In the construction industry, there are existing standards and codes that specify the quality standards of fire doors, or the manner in which concreting operations are expected to take place for high rise building projects.

There are also established measurement systems that specify what constitute good workmanship standards for building works. One example is the Construction Quality Assessment System. Top management commitment as well as employee involvement are two important ingredients for creating a comprehensive project quality management plan. The plan is expected to cover all the works involved in a project. By extension, quality is also to some extent related to environmental and safety issues. Hence, in large building projects, it is quite common to find a senior manager responsible for all three areas of quality, environment and safety. The integration and certification of these three areas can also be a prerequisite for the builder to register as a bidder for public sector building projects.

i. Planning for human resources

People are the most important assets in projects. It is people that make things happen. The work breakdown structure as well the network diagram provides the platform for the project manager to plan for when people are to be brought into the project. The project team may not want to do everything in-house. Some parts of the project may be subcontracted to external parties. This out-sourcing should also be factored into the human resource plan to ascertain if the external parties are available at the time when they are needed for the project. Some organizations may have direct labour that is pooled to be shared among projects. If there is a common labour pool, the project manager needs to ascertain when the labour pool is available for the project without upsetting the progress of other projects.

Where conflicts arise, the project manager should also mediate and resolve these conflicts in a timely manner. In some cases, the expertise for the work may not be available locally. For example, in the restoration of an old Chinese temple in Singapore, the artisans for the intricate wood carvings are found only in the Henan province of China. In such a situation, the project manager needs to make plans to bring these artisans to Singapore at the time when they are needed. Provisions also have to be made to house them properly when they are in Singapore. Based on the work breakdown structure and the network diagram, the project manager is able to know what, when and how many workers of a particular trade are needed for the project. It is also important to know the productivity rates of the various trades so that based on the quantities of work required the most appropriate numbers of tradesmen are brought to the worksite.

In the case of bricklaying, for example, it is necessary for the project manager to know how many bricks a team of four bricklayers can lay in one hour. It is often also not the case that more workers mean the tasks can be completed faster. Bringing more workers to the worksite may lead to congestion that can slow progress. It is also not a good practice to mobilize too many workers at the worksite for a short period of time that can lead to peaks and troughs in the project labour requirements. As far as possible, the human resource plan should factor in smoothing and levelling to avoid labour peaks and troughs. With many external parties working in the same worksite, the project manager should also ensure that there is proper co-ordination among them to facilitate smooth workflow. This is where effective communications come into place. The human resource plan is therefore closely allied with the communications plan.

j. Developing the communications plan

Communications provide instructions on how the project is to be carried out as well as report on the project progress. Poor communications can lead at best to hilarious outcomes and at worst to project failure. The development of the communications plan must therefore take into account the people who are involved with the project. The stakeholders can be both external and internal to an organization and may have a direct or indirect influence on the project. The communications plan ascertains who needs what information, when the information is needed and how the information is to be delivered. Such information can relate to reporting on project progress or the confirmation of deliveries of five ready mixed concrete trucks to the work-site on a specific day and time.

The communications plan should therefore include the identities and contact details of the various stakeholders involved with the project. In addition, some stakeholders are located overseas. The communications plans must therefore consider the different time zones as well as possible language and cultural barriers when information is transmitted. In the construction industry, communications modes can take many forms. These can include daily site meetings with the workers to provide instructions and weekly site meetings between the builder, client and the team of consultants to report on progress. Apart from site meetings, communications can be by means of telephone calls, email messages, tele-conferencing and formal letters. The various communications modes are by no means exhaustive. The most appropriate communications modes should essentially ensure that the correct information is sent to the correct people in a timely manner.

For mega construction projects, a dedicated project management information system may be set up for real-time communications to take place electronically and for records of decisions to be filed and maintained. Communications can both be brief and comprehensive depending on needs. Organizations should also include press briefings to the media and reports to shareholders within their communications plans. Project communications plans should also include clear information to be provided regularly to people living around the vicinity of the construction project to keep them updated of progress and upcoming inconveniences, if any.

k. Planning for risk management

Projects inevitably have risks of various types. These can be political risks, technical risks, organizational risks and environmental risks. Risk assessment takes into account the probability of their occurrence as well as the impact that these risks are likely to cause. Risks can be caused by nature such as the occurrence of earthquakes or typhoons. Risks can also be man-make such as those relating to poor hygiene leading to food poisoning. Some areas of works must have proper risk management systems in place as mandated by regulations such as those provided for in the Workplace Safety and Health Act. The work breakdown structure and network diagram provide the project manager with inputs on when, what and how some of these risks may occur.

It is important to recognize that mere understanding of what, when and how these risks may occur is not sufficient. What is more important is how the project manager takes measures to mitigate, reduce or eliminate these risks altogether. Project risks can come in various forms and magnitude. For mega construction projects involving significant payments to overseas vendors in their local currencies, the project manager is faced with foreign exchange risks. There are also risks associated with delays in the arrivals of imported building materials because of bad weather. At the construction site, safety infringements may lead to accidents and stop-work orders issued by the authorities. In project risk management, the project manager should plan early to prevent risk occurrence from turning into a crisis which may then spiral out of control.

There are avenues available for the project manager to manage risks. Measures that can be taken include risk mitigation by having more than one vendor for one type of building materials or by installing geotechnical monitoring devices to track excessive ground movement. Certain risks can be avoided by using another method of construction such as bored piles instead of drive-in piles to avoid excessive ground vibration. Risks can also be transferred to a third party through insurance. Finally, some risks may be low enough for these to be accepted by the project manager. Nevertheless, these perceptibly low risks must still be monitored to prevent them from deteriorating to become full-blown risks.

l. Planning for project procurement

Projects do consume resources. These resources can include labour, plant, equipment, materials and services. Through the work breakdown structure and the network diagram, the project manager is able to know what resources are needed for the project, how much of these resources are needed and when. Resources do not just appear out of nowhere. Procurement planning is required to ensure that all resources needed for the project are accounted for and are available. For mega construction projects, resources can come in many shapes and sizes even in the same category. For example, floor finishes can include carpet, vinyl tiles, ceramic tiles, granite tiles, slates and marble slabs. Some of these finishes are available locally while others must be imported from overseas. The project manager also needs to consider if the finishes can be purchased in bulk to enjoy better discounts and credit terms from the vendors.

The procedures for such purchases must also comply with organizational policies and practices. For example, public sector procurement would normally require open tender and at least three quotations to ensure that the monies expended are well spent for public accountability. The form of contract or agreement used for the procurement should also be considered as this can vary from the private sector to the public sector. The project manager needs to consider if some of the resources needed for the project are to be made in-house or purchased from an external vendor. Considerations for such procurement decisions include whether the project organization possesses the expertise necessary to manufacture these products in-house or if this can be done more economically by an external vendor who can benefits from economies of large scale production.

Apart from considering the resource types and placing the relevant orders, advanced notification and ample reminders have to be included in the procurement plan to trigger the vendor's attention. Hence, there is a strong link between the procurement plan and the communications plan. Deliveries of resources, especially building materials, should be well spaced out in the procurement plan to avoid site congestion. In confined sites where there is virtually no working space, the deliveries must be timed in such a manner that the materials so delivered can be installed immediately. Deliveries of related resources to the construction site should also be coordinated.

For example, in the concreting process, the project manager must ensure that the correct numbers of ready mixed concrete trucks arrive at the site at the appointed day and time. The trucks should not be dispatched too early to the site from the batching plant. If the trucks arrive much too early before the concrete is needed, setting will take place that may affect the concrete quality adversely. At the same time, the project manager needs to ensure that the concreting team is present to lay and cast the concrete. If lifting is required for the concrete at upper floors, the project manager also needs to ensure that the tower crane or the concrete pump is available and ready for use. All these resources do not simply converge on site at the same time by chance. The convergence is made possible only with a proper procurement plan in place. When procurement is completed with proper deliveries and payments made, the contract is then closed and filed for records.

m. Planning for project launch

When a collective agreement has been reached between the project manager, the project management team and the client, the project then officially commences. For significant and iconic building projects, the project launch may be commemorated by a ribbon-cutting ceremony attended by luminaries. In the case of a condominium project, the project launch may also serve to trigger marketing and sale activities. The project launch signifies to the project manager that all approvals necessary for work to commence on site have been obtained. These approvals are frequently issued by the authorities such as those relating to factory permits and registration of work sites. At this point in time, all the resources for the project should have already been identified with the corresponding quality, environmental and safety management systems put in place to ensure orderly construction. Appropriate resources are mobilized gradually as the project picks up steam.

In the case of construction, the project launch may be preceded by the demolition of any existing structures that are on the site. The site is then hoarded up for public safety. Temporary amenities such as site offices, canteens, stores, rest rooms, security posts and car parks are erected for use by the project team and the client's representatives. Utilities for water, gas and electricity supplies are connected for use. A site layout plan is prepared to ensure that all such amenities and utilities are properly located to achieve smooth work flow. With all these in place, the project launch physically takes off on site following the sequence of work identified in the network diagram.

3.5   The Executing Process Group

a. Executing follows planning

As noted above, planning is by far the most tedious process in project management. It is tedious because it needs to cater for all components, actions and activities required for project completion. In accounting for all these components, actions and activities in the work breakdown structure, planning also needs to factor in their associated risks as well as the time when appropriate resources have to be delivered to be used in the project. A well-executed project plan means half the job is already done. A comprehensive project plan should therefore cover all the resources needed for the project. The project plan also deals with the nine areas of the project management body of knowledge. It then remains for the project plan to be executed.

The project manager leads, directs and manages the project execution process, following the milestones identified and set out earlier in the project plan. Take for example the issues relating to quality in the project. In this case, the project manager needs to map the resources to the components, actions and activities to ensure that quality assurance is attained. If proper planning is already in place, together with proper execution, the outputs should be able to meet the specified quality standards. However, the output quality needs to be verified through inspections, reviews or audits. If poor quality is detected during execution, the project manager must stop the work in order to carry out corrective actions either through repairs or reworks. These have implications for additional costs and can also lead to time delays. This aspect of execution only relates to quality. The project manager must equally turn his attention to ensure that other areas are also executed properly. These other areas relate to scope, time, cost, human resource, communications, risk and integration for the project. Each of the nine areas in the project management body of knowledge is not looked at individually and separately.

These are all interconnected in one way or another as the project progresses through the work breakdown structure. A problem that surfaces in one area is bound to pose repercussions in other areas. It is therefore crucial for the project manager to put together a good team of people to assist him with trouble-shooting and to anticipate problems before they occur. The project management team helps by disseminating accurate information to all stakeholders in a timely manner. The team also manages procurement activities to ensure that the necessary resources are made available in the correct quality and quantities when these are needed. The key ingredients for successful project execution lie with having a comprehensive project plan in place and with a good team of people who works seamlessly together. The more detailed the project plan is, the more likely will execution be carried out successfully.

3.6   The Monitoring and Controlling Process Group

a. Need to monitor and control execution

While execution is more likely to succeed with good planning in place, things may still go wrong through no fault of the project manager or the project management team. Monitoring and controlling activities can be rendered more readily with the availability of a comprehensive project plan. As parts of the project plan are implemented or executed, monitoring and controlling activities also take place spontaneously for these parts. Hence, this is the reason why monitoring and controlling activities must still take place on a continuous basis to ensure that things happen the way they are supposed to as planned. While monitoring and controlling activities ensure that plans are carried out without a hitch, these activities also take into account the corrective actions that should be taken when the project is not proceeding as planned. It is important to recognize that monitoring and controlling activities are implemented to ensure that things go according to plans.

What is equally if not more important is that monitoring and controlling activities also take steps to fix mistakes immediately when these are spotted. Failure to take immediate rectification actions may lead to the mistakes or errors being irreparable or need more herculean efforts to rectify them at a later stage. Integrated change control management is an important part of project management as this allows the project manager to understand how one part of the project can affect other parts of the project. A project is made up of many interconnected parts. A change in one part may affect other parts. These interconnected parts not only cover the resources needed for the project but also cut across the nine areas of the project management body of knowledge.

Hence, for example, if there is a proposed change to the scope of the project relating to fire-rated doors, the proposed change is also likely to affect issues relating to the project time, cost, quality, risk, communications, human resource and procurement. Integrated change control must therefore take place to allow the project manager to understand how a change in the scope for fire-rated doors is likely to affect other parts of the project. The more significant and more frequent the proposed change, the more important it is for the project manager to exercise effective integrated change control management.

b. Areas to monitor and control in a project

A project is made up of many moving parts, all of which collectively contribute to the desired project outcomes. Monitoring and controlling cover not only these moving parts but also the ultimate project outcome. First and foremost, the project scope must be verified to ensure that only the parts necessary for the project are included in project execution. When completed, these parts must also be verified to ensure that the relevant stakeholder expectations have been met. This scope veri-fication exercise can be carried out using different approaches, including both formal and informal customer feedback. In the event that scope change occurs for valid reasons, monitoring and controlling must ensure that there is effective scope change control where all the affected stakeholders are informed of the change in a timely manner. Agreement for the scope change must be obtained and communi-cated to the affected stakeholders.

Monitoring and controlling also polices schedule control to ensure that all moving parts of the projects are completed as planned and that the critical paths in the network diagram are not unduly affected to lead to project completion delay. Among other enforcement tasks, monitoring and controlling is also extended to cost management, quality management, risk management and human resource man-agement with or without scope change. In the monitoring and controlling process, both external and internal stakeholders should also be managed. This is because stakeholders can influence or be influenced by the progress or slack of other stakeholders both upstream and downstream. Monitoring and controlling therefore ensure that timely stakeholder performance reports are tabled for evaluation by the project management team. The format of reporting and the relevant key perfor-mance indicators form part of the performance reports during the monitoring and controlling process.

Typical key performance indicators are related to costs, time and quality base-lines. When satisfactory performance reporting is completed monitoring and con-trolling cease for the related moving parts of the project. The closing process then takes place for these moving parts of the project. When performance reporting is not satisfactory, corrective actions may be needed to rectify defects. In such a scenario, monitoring and controlling take on an iterative mode back to planning and exe-cution. Change in project conditions may also lead to iteration as when a new risk is identified. The project management team then returns to the planning process group to deal with the new risk while at the same time, other project activities not affected by this change will continue to move on. This iteration continues until the required performance is achieved.

3.7 The Closing Process Group

a. Contractual and administrative closing

The work breakdown structure provides a list of components, actions and activities that must be completed for the project. In the construction industry, while parts of the work breakdown structure may be undertaken by the builder using direct labour, other parts of the work breakdown structure may be out-sourced to external vendors. In both instances, verification and auditing of the procurement should be completed satisfactorily before closing takes place. The closing process for the specific component, action or activity should consider the procurement documents to take into account the agreement that was earlier made. This is to ensure that what was agreed and contracted for earlier have all been fulfilled with respect to the quantities and the quality standards.

To facilitate verification and acceptance leading to closing, checks can be made with reference to relevant industry standards or codes of practice. In the construction industry, closing in some cases can only take place after testing and commissioning have been completed satisfactorily. Examples can include testing of the fire sprinkler system or commissioning of the electrical installation for a new shopping mall project. To ensure public health and safety, securing clearance from the fire safety bureau and a temporary occupation permit from the relevant building authorities may also be yardsticks to determine if closing can indeed take place for the respective vendors for the new shopping mall project. When all verifications, tests and commissioning exercises have been completed satisfactorily, the vendor contracts are then finalized and closed. This is when payments are made to the vendors after all contract terms have been fulfilled. Following the closing of the vendor contracts, administrative closing then takes place.

Administrative closing ensures that proper records of the vendor contracts are kept within the organization for a specific period of time. Formal letters to thank the vendors for their service and cooperation are also sent to officially end the working relationships. Feedback forms can be distributed to the various stakeholders to seek suggestions on how future improvements may be made. Stakeholder experiences are collated for lessons learned. Collectively, these are inputs to help build up the organizational assets for the firm to tap on for future projects. The project management team may also host an event where all stakeholders can come together to celebrate the successful completion of the project over a meal or some drinks. With this completed, the final task in the closing process is considered done.

3.8 Revision Questions

1. Why are the five project management processes important?
2. What are the reasons for the project management processes to overlap?
3. Which project management process is most important and why?
4. Which stakeholder first initiates a project and why?
5. What are the key considerations for initiating?
6. Why is planning the most tedious and time consuming process?
7. Do we need to start planning from scratch?
8. Do you agree that a project will never fail with good planning in place?
9. Why must planning in one area of the nine project management body of knowledge consider the other eight areas?
10. Is it true that perfect project planning cannot take place all the time?
11. Why is execution easy with good project plans in place?
12. Who are the external stakeholders in project execution?
13. What are the problems associated with executing too many activities at the same time?
14. Can projects live without monitoring and controlling?
15. What are some of the ways in which monitoring and controlling can be carried out?
16. What are some of the consequences arising from poor monitoring and controlling?
17. Are preventive actions as bad as corrective actions?
18. Is it true that poor planning can be rectified to some extent by good monitoring and controlling?
19. Is it true that poor planning lead to poor monitoring and controlling?
20. Is closing to be planned in?
21. Is the nature of closing different for each of the nine areas of the project management body of knowledge?
22. Can the termination of a contract be considered as part of closing?
23. What are some of the actions to be taken for closing to take place?
24. What are some of the actions to be taken after the closing process has taken place?

Chapter 4
Project Integration Management

4.1 Introduction

a. Why integration is important

Projects are made up of moving parts. As these parts evolve and move along over time, some parts will be completed on time while other parts may be delayed. Furthermore, some other parts may also face changes along the way. All these movements have to be co-ordinated and integrated so that all the various parts move in tandem towards successful project completion. In short, synchronization is essential. The moving parts of a project can come in many forms. Project integration management is one of the nine areas in the project management body of knowledge. As the project moves through the other areas relating to scope, time, cost, quality, human resource, communications, risk and procurement, project integration management is necessary to ensure that all knowledge areas are synchronized and that all the relevant stakeholders are kept informed of progress, delays and changes in a timely manner.

As each of the areas in the project management body of knowledge goes through the five project management processes, integration is also necessary to ensure that these processes keep in step with one another. In the five project management processes of initiating, planning, executing, monitoring and controlling as well as closing, the project manager needs to ensure that these processes are integrated seamlessly. If there is poor integration, monitoring and controlling may be triggered even before planning for an activity is completed. Project integration management serves to provide a platform where all stakeholders can readily tap on a common source of interconnected database or information for their use. Examples of such platforms include the Integrated Management System, Integrated Student Information Service platform and the Integrated Virtual Learning Environment platform.

If a central repository of related information is not maintained, much time will be wasted by the stakeholders to search for different types of information that they

© Springer Nature Singapore Pte Ltd. 2018
L.S. Pheng, *Project Management for the Built Environment*,
Management in the Built Environment,
https://doi.org/10.1007/978-981-10-6992-5_4

need as the project moves along. Furnishing and maintaining such a platform should therefore be part of the project manager's goals for project integration. This is especially critical in the construction industry where many stakeholders are involved even for a small building project. The resources needed for a construction project can span across manpower, materials, machines, money, management and methods. Each of these resources has their respective stakeholders who at some point in time in the project will need to work together. The stakeholders in the manpower category can include the clients, architects, quantity surveyors, engineers, contractors, suppliers and regulatory agencies. The materials used in the construction industry can include concrete, reinforcement bars, bricks, paints, tiles, doors, windows, pipes and cables. The machines used in the construction industry can include mobile cranes, excavators, generators, trucks, concrete pumps and welding equipment.

At some point in time, project financing will also involve the banks who provide loans to the construction firms as well as property purchasers who make progress payments to the developer when their homes are being built. Important project decisions are made by management staffs from the various stakeholder organizations who come together to meet, discuss and negotiate decisions that are perceived to be more beneficial to their own organizations. At the building site, the work processes are dictated by the various construction methods available for an activity such as concreting or internal partitioning. In the latter activity, the partition can be erected using the traditional bricks and mortar or the more productive dry-walls.

Hence, it can be seen that the numerous moving parts of a project must be integrated to ensure smooth workflow. If poor project integration management occurs, then these moving parts are likely to be disjointed to eventually fall apart. Beyond project management, integration is similarly critical for the program manager who needs to concurrently co-ordinate across numerous projects at the same time. Integration is also important for the portfolio manager to ensure that workflow across several portfolios of the same nature does not overtax existing resources unduly.

b. What project integration comprises

Project integration management is not a task that is completed in one day. This is an on-going venture that starts right from day one of the project. However seven important phases in which project integration takes place can be discerned from the beginning to the end of the project life cycle. These seven phases include developing the Project charter, Preliminary project scope statement, Project plan, Managing project execution, Monitoring and controlling the project, Managing integrated change control and Closing the project. Each of these seven phases should not be viewed in isolation.

As one phase is completed, the project manager moves on to the next phase with inputs provided from the preceding phase. For example, when the project charter is finalized, inputs are provided to the next phase for developing the preliminary project scope statement. Similar to the five project management processes, phases may also iterate as more information is made available to work out the details of the work breakdown structure. In addition, iteration may also happen among the phases because errors have been detected or requests for changes have been made. This iteration may happen between project execution and monitoring and controlling before closing.

4.2   Project Charter

a. What this document is about

For the project to progress with a common mission there must be consensus from all the major stakeholders. The project charter is an agreement that provides the guiding principles for the project team to move towards the common mission. The project charter shows who the major stakeholders are and their authority to act on behalf of their respective organizations. Prior to the development of the project charter and its specific contents, meetings and workshops can be conducted for the key stakeholders to deliberate and agree on major issues related to the project.

The project manager can serve as the facilitator in these meetings and workshops. Once the major issues have been agreed, the project charter is then drafted in the form of an agreement that sets out the key performance indicators expected of the project outcome. The key performance indicators can relate to time, cost, quality targets and environment goals. After a final review, the project charter is then formally signed by all the major stakeholders who thereby authorize the project as well as empower the project manager to procure resources to move the project forward. The project manager is required to accomplish successful project completion by achieving the stipulated key performance indicators.

b. What the project charter conveys

As an agreement, the project charter serves as a communication platform for all major stakeholders to arrive at a common understanding of how the project is to be executed as well as the desired project outcomes. Specific project requirements relating to the time, cost, quality, environment and other essential goals are set out in the project charter. All these requirements must be fulfilled at project completion. The larger picture of the project is also reflected in the project charter. For example, not all projects are profit-driven. If the project is for a social cause, this must be stated clearly in the project charter. The purpose of the project is also an important dimension that must be highlighted in the charter. If the purpose of the project is to explore feasibility or to exploit new opportunities, this should be stated clearly in the project charter.

In this context and with reference to the construction industry, an example of the considerations may include either rebuilding or upgrading of an existing building. Once a decision has been made, this must be stated unambiguously in the project charter. In addition, if the decision is to rebuild with the aim of embracing the latest information and communications technologies in the building, this should also be elaborated. The project charter also highlights the key milestones so that all stakeholders are clear about the project schedule with a target time for completion. It is also essential for the project charter to set out who the various internal and external stakeholders are who may influence the project outcomes. In the construction industry, the influencing stakeholders may include people living and working in the vicinities of the neighborhood. In this case, there may be existing regulations that require the project manager to keep in constant touch with the neighbors to update them with timely information of project schedule and progress.

This may take the form of regular group gatherings and meetings for the project team to answer queries and address the concerns of the neighbors. Where appropriate, the project manager can also make use of social media platforms to engage the external stakeholders identified in the project charter. Operating constraints with respect to external stakeholders should also be identified early in the project charter. This is particularly crucial in construction projects with limited working space and in close proximity to existing buildings. These constraints may include the need to curb excessive noise, dust, pollution, traffic congestion and threats to public safety. Promotion of good neighborliness should not be overlooked by the project manager. If some of these operating constraints are deemed to be noise-sensitive, for example, when the new construction project is next to an existing hospital or school, these constraints should already be identified and surfaced early in the project charter for further actions to be taken.

However, it should be acknowledged that not all information is available at this early stage of the project when the charter is formally agreed and signed. There are situations where the project team needs to make certain assumptions with respect to some of the operating constraints that the project is likely to meet in the near future. These assumptions may relate to the performance of the economy, market conditions, weather and availability of specialized materials or skills. The project charter may then spell out the options that can be taken in the face of these assumptions. The project charter also identifies the people who are directly involved with and responsible for the successful execution of the project from inception to handover. These people may come from different functional organizations such as the human resource department, legal department, accounts department and construction department. The key liaison person in each of these functional departments should be identified in the project charter. This then sets the scene for further elaboration in the near future when the project reaches a stage when project human resource management kicks in.

The project charter also highlights the procurement policies and procedures that deviate from normal practice. For example, if the developer client for a new mega building project also happens to operate a subsidiary company that supplies ready-mixed concrete, and has specifically indicated that concrete for the project must be procured from this subsidiary company, this directive should be made clear in the project charter. Finally, the project charter should also include an indication of the summary budget for the project. The summary should indicate the estimated budget, the amounts allocated to different parts of the project, and when the allocated budgets are expected to be spent.

4.3   Preliminary Project Scope Statement

a. What this statement is about

The preliminary project scope statement is the prelude to the development of the full scale project scope with supporting details. The preliminary project scope statement provides the framework for more details of the project to be solicited and made available as progressive elaboration takes place in the project. The statement provides the outline of what the project will create, achieve and deliver. It sets the scene for all the stakeholders for them to obtain a definitive and shared under-standing of the common purpose of the project. The preliminary project scope statement is triggered by the project manager and the project team by asking questions from the stakeholders on what they expect and desire out of the project. It also focuses on the corresponding amount of resources that the stakeholders are willing to commit in the project to realize their desired outcomes.

The statement starts with an idea or a concept. An example of a statement for a small building project may commence with the goal of constructing a playground for a village school in Cambodia costing $20,000 to be completed within one month funded by an international charity agency. With this as the starting point, the preliminary project scope statement is then approved by the project sponsor which is the international charity agency. As more details of what to include in the school playground project become available, the preliminary project scope statement will be refined through further planning and approval from the project sponsor. It should be noted that the preliminary project scope statement is not a mere one-paragraph statement. Depending on the size and complexity of the project, the preliminary project scope statement can end up as a modest, stand-alone document.

b. Developing the preliminary project scope statement

The preliminary project scope statement builds on the consensus agreed earlier by all stakeholders in the project charter. While the project charter provides the framework to guide the project manager and the project team, more details are progressively elaborated and developed through the preliminary project scope statement following the signing of the charter agreement. Depending on the size and complexity of the project, many or few items can be included in the statement. Nevertheless, the statement should include key dimensions relating to the objectives and deliverables of the project. Take for example the construction of a new government-funded sports stadium in an existing public housing estate.

The objective of the project sponsor can be stated to be for the promotion of a healthy and active lifestyle in the community through the sports stadium project. The deliverables for the sports stadium can include running tracks, basketball courts, badminton courts, swimming pools and rock climbing walls. Some of the project characteristics can include the specifications of the running tracks to be 400 metres long and the swimming pool to be at least 1 m deep. The project boundaries should also be spelled out in the preliminary project scope statement. For example, if the rock climbing walls exclude safety harnesses personalized for the individual user upon project completion, this exclusion should be specified in the statement. Likewise if future maintenance of the stadium is not included in the project scope, this should also be highlighted in the preliminary project scope statement. The identification of the project deliverables and characteristics should be supported by their corresponding acceptance criteria. For example, in the construction of the new swimming pools, there are certain industry's codes of practice that the pools should comply with. The prevailing regulations or standards may also prohibit the use of glazed ceramic tiles on floors near the vicinity of the swimming pools to prevent users from slipping over wet surfaces.

The key acceptance criteria for all the project deliverables should be identified in the statement, to be further elaborated progressively as more details are available. The assumptions that have been highlighted by the stakeholders earlier in the project charter are now expanded in the statement to elicit the constraints that the project is likely to face when working around these assumptions. An example of such constraints may relate to the availability of certain specialized building materials that can only be sourced from overseas. Such constraints may correspondingly lead to risks being identified from the assumptions made. Hence, the statement should include a list of initial project risks that are highlighted early to be dealt with as project planning intensifies. The unavailability of the specialized building materials at the time when they are needed is a project risk. These risks are also mapped against a time-line to estimate when they are likely to occur and when mitigation measures need to be taken.

The time-line of the likely risk occurrence is in turn mapped against the esti-mated schedule that shows the key milestones of the project. In the example of the new sports stadium project, the client may specify a deadline for completion and handover. With knowledge of this deadline, the project manager then works backward to estimate when major segments of the projects such as the running tracks, swimming pools and basketball courts are to be completed. The statement will then elaborate on the initial work breakdown structure for each of these major segments of work. If the segment of work relates to the basketball courts, then the initial work breakdown structure for the basketball courts are further developed to include the various relevant components, actions and activities. Similar initial work breakdown structures are also developed for other major segments of the project, bearing in mind at the same time of the need to integrate the various initial work breakdown structures to seamlessly contribute to the common goal of successful project completion. With the initial work breakdown structures completed, the rough cost estimates for the components can then be computed for inclusion in the preliminary project scope statement.

These rough cost estimates also take into consideration management and staffing requirements as well as their costings for the project. The statement should also indicate how approvals for the project are to be processed. Special approvals that are needed for the project should be highlighted in the statement. Examples of projects that require special approvals from the authorities include the construction of casinos and high-rise buildings that are near airports. The statement should also indicate clearly when and from whom approvals should be routed if there are proposed changes in the near future when the project moves along. The preliminary project scope statement therefore contains a suite of information and instructions which certainly do not end up as a mere one-paragraph statement. This statement is important for integration because it sets the groundwork for further progressive elaboration when full-scale project scope management kicks in.

4.4 Project Plan

a. What the project plan is about

Following the agreement of stakeholders through the project charter and completion of the preliminary project scope statement, the project plan can now be developed with more certainty. Nevertheless, at the early days of the project, the plan is still sketchy and details as well as decisions need to be made by the stakeholders. This is where the project plan also serves as an integration platform to ensure that all stakeholders are kept informed as and when details are made available and decisions taken. The project plan guides the stakeholders on how the project should progress and how it should be managed. For example, in the construction of a detached house, various resources such as materials, equipment and manpower are needed. These resources are used for different parts of the house and can include the foundations, structural works, architectural works, mechanical and electrical services as well as external landscaping.

These resources do not just appear out of nowhere. The project manager needs to identify what resources are needed, in what quantities and when these are needed as the project progresses. In the case of building materials, the project manager will need to identify the different types of materials such as concrete, reinforcement bars, bricks, doors, windows, floor finishes, electrical cablings, etc. To ensure that these building materials are delivered and installed at the building site, the project manager needs to plan for each of these building materials using a project plan. The project plan is in turn made up of a bunch of plans that cover scope, cost, time, quality, human resource, risks, communications and procurement for all and sundries needed for the house project. When each of these plans are developed and implemented, the five project management processes will be triggered. For example, in the case of the quality management plan, implementation will include the following processes of initiating, planning, executing, monitoring and controlling, as well as closing. In summary, the project plan as a whole reflects the value systems, beliefs, priorities and conditions of the project.

The project plan is an integration platform that brings all the stakeholders together. It communicates to all the stakeholders and the project management team how the project is to be run, managed and controlled. For a project plan to be realistic, it has to cover all the foreseeable components, actions and activities needed for the project. The project plan can therefore be a massive document. It serves to provide a structure for the project; for example, in the case of the detached house project, by annotating the substructure works, superstructure works, architecture works and building services. Through the detailed annotations, a set of documents is created for reference and use by the project management team. These documents then help to facilitate clear communications between all stakeholders. Examples of such documents can include specifications and extracts from relevant codes of practice. These then provide the baselines or quality standards to form the acceptance criteria to assess various parts of the project when these are completed. For example, fire doors should be installed with push bars to be opened outwards for escape purposes.

b. How the project plan is developed

Project plan development requires a constant process of progressive elaboration which follows a logical and systematic approach. If a similar project has been completed earlier, the project manager may refer to the project plan of the completed project for the lessons learned. The template used in the project completed earlier may also be tweaked and adopted for use. Otherwise, the project manager will need to start with a clean slate if it is an entirely new project with no precedence. If this is indeed the case for an entirely new building project using innovative technologies for the first time, then the project manager will first need to review the business requirements, functional requirements and technical requirements of the new proposed facilities.

A review of the business requirements will lead to identification of the desirable functional requirements, which will in turn, lead to identification of the corresponding technical requirements. The review may not be a straightforward process for large and complex projects where numerous iterations are to be expected before decisions are made. A design plan then evolves following the completion of this review. The design plan provides the platform for the work breakdown structure to be developed for the project. The work breakdown structure decomposes the entire project into manageable bits where further planning can then take place. For example, the work breakdown structure for substructure works can include deep basement excavation, diaphragm walls, piling, ground beams, basement construction, waterproofing and dewatering of groundwater.

With the manageable bits identified, the estimated time needed for their completion is then established after considering and selecting the resources needed for their execution. The project schedule can then be developed. This then moves to final planning with more details to be provided progressively for all components, actions and activities in the work breakdown structure, covering the various domains of the project management body of knowledge such as quality, risks, communications, procurement, etc. In the progressive elaboration process, changes may occur. These changes then iterate back to revise the work breakdown structure for the developmental cycle to be repeated until there are no further changes to the project plan. These iterations with inputs from relevant stakeholders also serve to promote project integration management.

c. The bunch of plans

The project plan consists of a bunch of plans, each with their own focus area. This allows planning to take place in a more manageable bite-size and concentrated manner. For example, in the construction of a new condominium project, many issues have to be addressed across different components, actions and activities in the work breakdown structure. The bunch of plans helps the project manager to focus on one area at any one time. For example, there is a plan for quality management, another plan for communications management, and yet another plan for risk management and so on. While the various plans are undertaken to focus on only one area, these plans need to be integrated concurrently at the same time by the project manager. This is because a decision made in one plan can affect decisions in other plans for the same components, actions and activities.

For example, in the quality management plan for fire safety doors, issues relating to their corresponding costs, communications and human resource requirements must also be annotated at the same time. In summary, the bunch of plans that make up the overall project plan should include the following items: scope management plan, schedule management plan, cost management plan, quality management plan, human resource management plan, communications management plan, risk management plan and procurement management plan. This is in addition to the integration management plan that serves to ensure co-ordination across the board. Based on this bunch of plans, the project manager is better positioned to seek out areas for process improvements, establish the list of milestones (for example, when foundation works will be completed), determine the resource calendar (for example, when certain specialist trades are needed on site) as well as frame the risk register (for example, identify the nature of risks for deep basement excavation). In addition, the bunch of plans will provide the project manager with the baselines to compare actual performance with planned performance.

These project baselines allow the project manager to assess if the project is ahead or behind schedule and if there is overspending or under-spending of the planned budget. While the overall project plan evolves and takes shape, it should also be acknowledged that not all decisions can be made at this early stage of the project when some information is still unavailable for planning to take place. At this early stage, open issues or unresolved matters may still be pending that require further resolution as project planning progresses. For example, the project manager may need to await further decisions from the client for additional financing from the banks or await further confirmation from an overseas vendor on the availability of specialized plant and equipment. Nonetheless, developing the overall project plan and its attendant bunch of plans offers all stakeholders the opportunities to integrate their discussions to reach common decisions in a clear manner.

4.5 The Project Plan in Action

a. Executing the bunch of plans

The outcome of the overall project plan is the bunch of plans that is now ready for implementation or execution. At the execution stage, components, actions and activities identified from the work breakdown structure are systematically carried out according to plans. The bunch of plans therefore provides the platform for integration as different stakeholders come together to provide their individual contributions. The project manager uses these plans to co-ordinate progress of the work to, among other things ensure that activities are not carried out too early or too late. The bunch of plans ensures that all the work completed do satisfy the project objectives set by the client in an integrated manner.

These objectives are set out in the scope management plan. The cost management plan ensures that funds are appropriately spent to meet these objectives. The human resource management plan maps the directions for the project manager to lead, manage and train members of his project team. The procurement management plan provides the framework for acquiring, using and managing various resources needed to complete all components, actions and activities identified from the work breakdown structure. After resources are delivered to the project and used, the plan also outlines the procedures for completing the necessary procurement requirements for closing. The risk management plan identifies the risks as well as the persons responsible for managing these risks as and when they occur.

All these operations require clear communications, instructions and directions which are surfaced in the communications management plan. As the project progresses, the time management plan, cost management plan and quality management plan kick into collect data and information on performance reporting. Through these plans, the project manager is able to monitor if the project is on schedule, if there has been overspending and if the quality standards delivered are acceptable. In the event where changes happen, the bunch of plans should also surface the procedures for these changes to be approved by the appropriate authorities.

After approvals for changes have been obtained, relevant portions in the bunch of plans will be updated to reflect the effects of these changes on the project. For example, changes may occur when additional funds are made available to the client in a new private school building project. The client may then decide to incorporate further features such as a garden and swimming pool on the roof top of the new building. Immediate changes are therefore made to the scope management plan. These in turn have a snowballing effect on the other plans affecting cost, time, quality, human resource, risk, communications and procurement. The bunch of plans is therefore synchronized to accommodate these changes. In the process, integration is achieved through the important process of integrated change control.

This integrated change control process also identifies where corrective actions and preventive actions have to be taken if necessary. Apart from managing the change requests, the integration management plan also provides directions for rectifying defects and managing repairs. Finally, the bunch of plans should also make provisions for documenting the lessons learned when the project is completed. This documentation is an important part of organizational assets for future lessons when similar projects are undertaken. Documenting the lessons learned is a good habit because this allows good practices to be carried forward to the next project. It also provides warnings on what should be avoided in the next project.

b. Monitoring and controlling through the bunch of plans

The bunch of plans provides the platform for all components, actions and activities of the project identified through the work breakdown structure to be carried out smoothly. This is the implementation or execution part of the bunch of project plans. In addition, apart from executing the work in a systematic manner, the bunch of plans also serves to provide the various tools and techniques used to monitor and control the workflow in the project. The bunch of plans therefore serves two purposes; namely for execution as well as for monitoring and controlling. While execution of work is carried out systematically using the bunch of plans, these plans also facilitate systematic monitoring and controlling to take place. For smaller projects, monitoring and controlling may take place manually.

However, for larger and more complex projects, many workflows and activities may be linked and a change in one activity may affect another activity elsewhere in the project. In such a situation, a project management information system is useful for monitoring and controlling. The system allows the procedures to be automated and along the way, prompts questions for the project manager to take timely actions without overlooking the details. Importantly, these questions seek answers to time and cost performance of the project. This is often done through earned value management which comprises a set of formulas that measure and report on budget and schedule progress.

Through earned value management, the project manager is able to assess if funds are being spent faster than what the project schedule has allowed for and vice versa. Follow-up actions must then be taken to ensure that the project cash flow and schedule remain healthy. The trade-off between spending a bit more to obtain better returns can be considered. In some instances, monitoring and controlling is also undertaken through consultations with the experts. These can be third party specialist consultants, project team members and individuals who possess the domain knowledge to offer expert judgment on the subject matter. For example, in the case of basement construction involving deep excavation, a geotechnical engineer may be consulted to monitor and control safety on site as the work progresses. In the final analysis, monitoring and controlling the project means examining the outcomes of the work that is both on-going and completed. These may include recommendations for corrective works, recommendations for preventive works, recommendations for repair works, change requests as well as projections based on earned value management tools.

Hence, there are many things that the project manager needs to juggle in the monitoring and controlling process. To facilitate this process, integrated change control is recommended. With this set up, all change requests must pass through integrated change control for assessment, approval, dissemination of information for follow-up actions and keeping records. For large and complex projects, a real-time on-line integrated change control system is recommended to be part of the project management information system. Whenever changes are proposed, the change requests are directed to this change control system. The request is then routed to a designated change control committee who then evaluates the merits or otherwise of the proposed change request. The committee applies a set of questions that query the value of the propose change, the cost-benefit analysis of the proposed change, the risks involved if the proposed change is approved or rejected, and the socio-political implications arising from the change request decision.

If there are already established procedures in place in the project to pre-approve certain change requests, then these procedures will be applied. For example, if the project stakeholders have earlier agreed that change requests that involve less than $20,000 can be approved by the project manager, then there is no necessity for this change request to be subject to review by the change control committee. Such a request should be approved readily but should still be channeled through the change control system for information dissemination and records keeping. Otherwise, for change requests that are more than $20,000 or more, the requests must be routed to the change control committee who either approves or disapproves the requests.

If approval is granted, then all the affected project baselines must be updated and the corresponding parts of the overall project plan must also be amended for all stakeholders to be kept aware of such a change approval. If the change request is not approved, then the reasons for the rejection must be conveyed in a timely manner to the stakeholders concerned. The decision for the rejection, the reasons as well as all supporting documents must then be archived for records purposes. The archival of the decision, the reasons and the supporting documents is an important step to avert any disputes that may arise in the future, especially in the closing phase of a project. These steps also collectively help to ensure integration management in the project.

4.6   The Closing Process

a. What closing the project entails

When all works are completed, the project naturally comes to a close. This is the stage where the stakeholders are satisfied with the project quality outcomes and when the vendors are then paid for any outstanding monies. It should be noted that there may be more than one closing in a project. This is particularly true in large and complex projects that are carried out in phases. If the project life cycle has multiple phases, then there will be a closing process for each phase. For example, a large integrated resort project may comprise of hotels, casino, theme park and infrastructure works. The project starts with and completes the infrastructure works first before other building works commence. The closing process for the infrastructure works is therefore in a different phase relative to the other building phases.

Similarly, a large condominium project with more than 1,000 apartment units may be developed in phases, with each phase completed and marketed at different times. Following completion of the project or phase, administrative closure then takes place. This administrative task is to among other things ensures that proper records are kept for future references. In administrative closure, all pertinent project records are collated, assembled and stored in an appropriate location with proper referencing for easy retrieval. During administrative closure, the project manager also analyses which aspects of the project are deemed to have succeeded or failed according to the baselines or metrics that have been set for their evaluation. Lessons that correspond to these successes and failures are then written and compiled by the project manager.

These lessons will form part of the organizational assets that provide clear documentation of what to duplicate and what to avoid in future similar projects. Collectively, all such project documents are then archived in a safe and readily accessible location for future references. At the broader level of the project, there is also contractual closure, which also encompasses administrative closure. Hence, administrative closure is part of contractual closure. The latter however deals primarily with agreements or contracts that the project organization signed with the vendors and the client. Following the completion of the project, the contract or agreement is deemed to have been fulfilled satisfactorily. Contractual closure also signifies the completion of the product verification process with proper documents for the works that have been accepted.

Contractual closure also includes documents relating to invoices and payments, with proper receipts and acknowledgements duly filed for records purposes. Such documents should be filed together with their respective formal offer and acceptance documentations that have been administered earlier. In cases where there have been early contract terminations because of vendor delinquencies, these should be annotated as such with the termination conditions clearly highlighted for future references. Where there have been problematic cases arising from variations or delayed payments, these should also be recorded for contractual closure when all issues have been resolved. All such documents, including commissioning and operations manuals, should be filed in properly catalogued and clearly labelled project files that form the project closure documents. These documents will also serve as historical documents for future learning purposes.

4.7 Synthesis

a. Moving together in a synchronized manner

Project integration management starts from day 1 of a project. It starts with the co-ordination of what little information exists at the early stages of a project. As more information is made available as the project progresses, integration management takes on a wider role as it cuts across the various areas of the project management body of knowledge as well as the project management processes. Integration management starts with and includes the project charter, the preliminary project scope statement, monitoring and controlling as well as closing. It relies primarily on project plan development, project plan execution and integrated change control to achieve proper co-ordination.

All moving parts of a project are managed through an integrated change control system, with as complete information as possible to be disseminated to all relevant stakeholders. Based on the work breakdown structure, all project components, actions and activities have to move together in a synchronized manner. Integration management is also about negotiation because there will inevitably be undercurrents from competing project objectives. Apart from project management and leadership skills, integration management is also about general management skills that call for effective planning, organizing, co-ordination, controlling and clear communications among all stakeholders.

4.8 Revision Questions

1. Is project integration the same as co-ordination?
2. Why is there a need for integration for both the project management body of knowledge and the project management processes?
3. Is integration required in program management?
4. Is integration required in portfolio management?
5. Who are the stakeholders in a typical construction project?
6. What are the resources needed for the construction of a typical building project?
7. Is project integration management a day-to-day routine?
8. What are the moving parts of a project that require integration?
9. Why is the project charter an important document to facilitate project integration management?
10. Who are the stakeholders involved with the project charter?
11. How is a project charter developed?
12. What is the important information set out in the project charter?
13. What needs to be done for stakeholders to agree to the project charter?
14. What are the benefits for stakeholders leading to the project charter agreement?
15. Why is the project charter important for integration?
16. Why is the preliminary project scope statement an important document at the early stages of project integration management?
17. Which stakeholders provide inputs to develop the preliminary project scope statement?
18. Is it true that the preliminary project scope statement is just merely a short statement?
19. What are the key items covered in the preliminary project scope statement?

20. How is the preliminary project scope statement relevant for integration management?
21. Why is it necessary to commence formulation of the project plan only upon the completion of the preliminary project scope statement?
22. What does the project plan cover?
23. Is it necessary for the development of the project plan to follow a logical and systematic approach?
24. Why is the project plan essential for integration management?
25. What make up the "bunch of plans" in the project plan?
26. Why are baselines important for integration management in project plans?
27. What does execution of the project plans entail?
28. Following the completion of project plan execution, why is it necessary to create "lessons learned documentation"?
29. Are preventive actions better than corrective actions in project plan execution?
30. Why is a real-time project management information system useful for monitoring and controlling the project work?
31. What are some of the possible outcomes that follow an assessment of the results of completed project works?
32. Why is integrated change control an important process for projects?
33. What constitutes administrative closure for projects?
34. What constitutes contractual closure for projects?

Chapter 5
Project Scope Management

5.1 Introduction

a. Three major components in scope management

Project scope management is closely aligned with integrated change control. In essence, a project needs a scope statement to steer the project in the right direction. But the scope statement is not a simple one-paragraph statement. Depending on the scale and complexity of the project, a fully developed project scope statement can be a set of hefty documents. These documents can include elemental breakdown of the entire building project, specifications and drawings. Based on progressive elaboration, development of the project scope statement would require many inputs from stakeholders to define the scope of work clearly and without ambiguities. There are three major components in scope management. These include establishing the scope, managing scope change, and verifying scope.

5.2 Establishing the Scope

a. Purposes of project scope management

The project manager needs to focus on what is needed to complete the project successfully. To secure completion, resources are inevitably expended. Hence, the project manager requires a definitive project scope statement to focus the efforts needed to rally all the resources necessary for the project. With this focus in place, organization and concentration can then take place. Fundamentally, project scope management ensures that only the works necessary to complete the project are undertaken; no more and no less. To stretch this further, for example, McDonald's only sells burgers; it does not sell cars. Likewise, Toyota only sells cars; it does not sell burgers. This sums up the essence of project scope management.

© Springer Nature Singapore Pte Ltd. 2018
L.S. Pheng, *Project Management for the Built Environment*,
Management in the Built Environment,
https://doi.org/10.1007/978-981-10-6992-5_5

As one of the key areas in the project management body of knowledge, scope management can be defined as the project management processes that ensure only the works necessary for the successful completion of a project are undertaken. To ensure that this happens, a project scope statement is therefore necessary. Apart from defining what is included and excluded in the project, the scope statement also describes all the work necessary for successful delivery of the product. It is important to first establish the project objectives before work commences on formulating and developing the scope statement. The resources needed to achieve the project objectives are then identified through the scope statement in no uncertain terms. The scope statement provides the point of reference or compass to guide the project team to do only what is necessary for the project. In the final analysis, if the project objectives seek to construct a school building, the project team should not end up erecting a bridge!

b. What the product scope entails

The project scope statement identifies the deliverables for the project. The deliverables are in turn made up of products. Hence, apart from the project scope, it is also necessary to describe the product scope at the micro level of the project. The project scope defines the work necessary to create the various deliverables of a project. For example, the refurbishment of an existing school building constitutes the overall project. Part of the refurbishment project may include creating information and communication technology-enabled classrooms. These classrooms with the latest smart technologies incorporated will be some of the key deliverables of this refurbishment project. Within these smart classrooms, many latest information and communication technology related gadgets will be installed. These smart gadgets and computer systems are the products in one of the deliverables in the school refurbishment project. This simply means that there are three levels of analysis for scope management. At the highest level, the project scope (i.e. the school refurbishment project) determines the work necessary to produce the deliverables. At the second level, the deliverables (i.e. the smart classrooms) are described. At the third level, the products (smart gadgets and computer systems) are similarly described.

Hence, the description in the product scope statement is an indication of the characteristics and attributes to be found in the deliverables that the project is creating. By extension, there are linkages that bind the project scope and the product scope. The product scope statement describes the features, attributes and characteristics of the product that is to be installed in the project; for example, real-time globally connected smart classrooms. It is necessary for the project manager to first understand the stakeholders' requirements for these to be incorporated in the product scope. The product scope is then evaluated against these requirements. At the broader level, the project scope is evaluated against the bunch of plans that make up the overall project plan. The "bunch of plans" includes the project quality management plan, project risk management plan, project communications management plan, etc.

Tied closely to product scope is the process of product analysis which requires the project manager to understand all the features and characteristics of a product, what it can do and cannot do. The project manager is able to conduct product analysis in several ways. Firstly, the product can be broken down progressively into smaller parts or components for evaluation. Hence, in the school refurbishment project, the project manager needs to understand what the products likely to be used in various spaces of the school are, such as the smart classrooms, the indoor sports hall and the pastoral care facilities. The products in each of these spaces are then identified and their capabilities and features analyzed to determine if these are suitable for their intended use in the spaces concerned. For example, there can be a rock-climbing wall in the indoor sports hall. The product analysis for the rock-climbing wall will, among other things, evaluate if the safety harnesses can be securely fixed and comfortably used when installed.

This is where a good understanding of the client's needs relating to rock-climbing as a sporting activity within the school is essential. This understanding must be holistic. Systems engineering can be adopted by the project manager to study the issue at hand from a systemic perspective. This is to ensure that the product to be installed in the project is able to meet all the client's needs relating to costs, quality, safety and other requirements. In many cases, product analysis will also highlight competing choices that are available for the client's selection. This is where value management can be adopted by the project team to ensure that the best interests of the client can be achieved. In this context, due consideration should be given to balancing quality and costs in achieving the best or optimal value for the client. As part of value management, value engineering facilitates assessments to be conducted to reduce costs and yet at the same time to correspondingly enhance client's satisfaction or profitability.

Value analysis can also be undertaken to ensure that accepted costs tally well against expected quality demanded by the client to achieve good value for money. As part of product analysis, further evaluation may also need to be undertaken as part of function analysis to complement value engineering and value analysis. In function analysis, the project manager searches for the best rational solution to a problem through testing the functions of a specific product, graphing the results, understanding the trends and then making a recommendation to the client. For example, function analysis can evaluate the durability of a door lockset in a busy public space to determine its expected wear and tear through constant use over time. How the evaluation of the door lockset will turn out to be can also be influenced by its quality function expected by the client. Hence, it is paramount for the project manager to build in customer needs when undertaking product analysis.

c. How the project scope is planned

The project scope is not normally rendered completely definitive in a single day. This is particularly true of large complex building projects where many stakeholders are involved who have to make decisions relating to competing choices. Generally speaking, iterations are involved with the project scope being defined broadly at first and through progressive elaboration, for refinements to take place eventually. Through these iterations, progressive elaboration and refinements, planning for the project scope is focused only on the work required to create the products necessary for the project, no more and no less. Hence, the project scope work and the product scope features are closely related to one another.

Following the completion of project scope planning, a project scope statement and the project scope management plan are created. To reiterate, the project scope statement describes what the project is all about. The project scope management plan allows the project manager to work hand in hand with the five project management processes of initiating, planning, executing, monitoring and controlling as well as closing. For example, the project scope management plan for the school refurbishment project may include the installation of 25 new fire rated doors. After the project has started, the project scope management plan alerts the project manager to initiate actions to procure the new doors that meet the specified quality, plan for when these doors should be delivered to the building site and who, how and when these new doors should be installed. For quality control, the project manager then monitors and controls the door installation. When all doors have been installed as specified, this part of the school refurbishment project is then closed.

d. How the scope management plan is created

The project manager needs not reinvent the wheel to create the scope management plan. There are tools which the project manager can make use of for this purpose. Firstly, the project manager can rely on expert judgement to create the scope management plan. This is particularly useful for new projects which the project manager or the project organization do not possess the necessary know-how. For example, the new projects may include the construction of a new flyer, an integrated resort or dirt biking race track. In such cases, building professionals who have had past experience of such projects can be consulted to seek their inputs to develop the scope management plan. Secondly, and more commonly the case, the project manager can rely on existing templates, forms, standards and other organizational assets for guidance in developing the scope management plan.

When completed, the scope management plan provides guidance to the project team on how the necessary scope of work is to be managed, and how changes or variations, if any, are to be included in the project plan. The scope management plan also provides guidance on how likely scope changes are to be expected when the project starts, and how much the scope is allowed to change. Generally speaking, the likelihood of scope change is higher during the early stages of a project. As the project progresses, it can become more costly to incorporate further changes to a project. Hence, as the project moves towards completion, the likelihood of changes is lessened. Major changes should not be expected after the project passes the design freeze phase. The scope management plan also provides guidance on the costs aspects relating to changes and the limits within which changes are disallowed or allowed with exceptional reasons. Granted that changes are to be expected in large complex building projects, the scope management plan plays an important role in defining four areas of work.

Firstly, the scope management plan provides guidance on the process that the project team and stakeholders are likely to go through in creating a detailed project scope statement. Such guidance is also likely to be tied to a stipulated time-line, such as when the design freeze is expected to be reached. Without such a stipulated time-line, discussions can go on forever without reaching a conclusion. Hence, the scope management plan informs all parties of the various deadlines when important milestones and decisions have to be made. The scope management plan also provides initial guidance on the nature and type of work that the project includes. This can range from the foundation works at the ground level to the topping out works at the roof level of a high rise building.

Secondly, with the above completed, the scope management plan provides guidance on creating the follow-up work breakdown structure. The work breakdown structure sets out in greater details the numerous components, actions and activities that the project includes. For example, the foundation works necessary for the building project are further elaborated to include supporting details such as excavation, dewatering, disposal of excavated soil offsite, temporary supports to sides of excavation, piling, basement construction, etc. In typical building projects, the scope management plan may include the following major work items as part of the work breakdown structure: namely, substructure works, superstructure works, architectural works, mechanical and electrical works, and external works.

Correspondingly, and thirdly, the scope management plan also highlights the process required for checking and formally accepting the various project deliverables by the client. Such checks can include walkthroughs of the building completed or confirmation of laboratory test results. These checks are essential for closing the parts of the project that have been completed satisfactorily as described in the scope management plan.

Fourthly, the scope management plan also provides guidance on how change requests are to be submitted and assessed. For example, the client may specify that change requests costing less than $20,000 can be approved by the project manager. However, the client may also instruct that any change requests that are likely to add more than $20,000 to the project budget must be submitted to the client's organization for a review committee to evaluate. The review committee can decide to either approve or reject the change request. It should therefore be noted that the scope management plan does not simply spell out what is to be included or excluded in the project. It goes beyond this to include quality checks, formal acceptance as well as change requests.

e. Scope definition

Defining the scope clearly is important for creating the scope management plan. It is crucial to get the scope definition right because this sets the platform for including only the works necessary for the project. The scope management plan may consist of several scope definitions for different stages of the project; namely what does each project stage includes. Essentially, the task is to break a large project down into chunks that are more manageable for the human mind to comprehend meaningfully. In the case of a building project, such a task would normally follow a rational and logical approach to designing, planning, constructing and commissioning the building. The major stages of a building project within which the scope definition may take place can include the conceptual design stage, schematic design stage, detailed design stage, preparation of construction documents stage, tender stage, construction stage, as well as commissioning and handover stage.

This logical approach is useful because at the early stages of a proposed project, most stakeholders would not know what they want even though they may have a general idea of what the project is all about. For example, in a large integrated resort project involving theme parks, casinos and hotels, detailed planning is necessary for the project management team to pin down exactly what are involved in the project. Each of the theme parks, casinos and hotels must therefore be furnished with appropriate scope definitions of their different parts that would eventually contribute to the creation of the overall scope management plan. In summary, detailed and fully developed plans are required for the project management team to progress towards successful project completion.

f. Inputs to define scope

Scope definitions as well as the scope management plan do not simply materialize out of nowhere. Someone with the necessary know-how must provide the inputs to create the scope definitions and the scope management plan. The first available resource to fall back on to garner inputs for this purpose would be the organizational assets or records that the project manager has access to. These assets are normally available because similar projects have been completed in the past by the project organization. These organizational assets can be both formal and informal documents that set out the policies, guidelines and procedures for the project manager. For example, because of past client's instructions, there can be guidelines relating to what the project manager can and cannot procure resources from and if allowed, how such resources are to be procured. The constraints on procurement of resources can have implications for project scope definition as well as in further developing the scope management plan.

If there are past and current client's instructions that limit the boundaries of what the project can and cannot do, then these should already be reflected earlier in the project charter and agreed upon by all the stakeholders. The project charter is therefore another input for scope definition. Following this, the preliminary project scope statement is then developed. It is worth reiterating that the preliminary project scope statement is not necessarily a simple one-paragraph statement. Depending on the nature, size and complexity of the project, the preliminary project scope statement can turn out to be a sizeable document that includes sketches and drawings for clarity. When fully developed, the project scope management plan provides yet another document for scope definition. The project scope management plan spells out what changes are allowed and to whom such change requests are to be routed to for approval. Hence, the procedures to seek approval for change requests are tied closely to the project scope management plan. Collectively, these form the inputs for and can therefore correspondingly influence the scope definition of the project.

g. Other activities to creating scope definition

Scope definition can be rendered onerous to the project manager at the early stages of a project because there can be many choices and alternatives available for client selection. To provide true value to the client, the project manager should start with looking out for alternative solutions that meet the project's needs. For example, the project manager should source for and examine more than one type of information technology product as well as several vendors to assess which solution best meet the client's needs. This assessment should be conducted with an understanding of what the project stakeholders expect. In this context, the voice of the customer is therefore an important input when performing stakeholder analysis.

A balance needs to be maintained between what the vendors can offer and what the project stakeholders need. Once a decision has been made on the selection of an alternative solution over an earlier proposed solution, the affected stakeholders should be informed with justifications provided for why the change is being made. Once the alternative solution has been approved for adoption, the scope definition will need to be updated with the necessary supporting details. This continues until all tasks associated with scope planning are completed, leading eventually to finalization of the scope statement. For large complex projects, the scope statement is clearly not a simple one-paragraph statement. The scope statement can be a hefty document that provides guidance for all project decisions in the future when change management takes place to account for variations.

As a hefty document, the scope statement provides details relating to the project objectives, product scope description, project requirements, project boundaries, project deliverables, project acceptance criteria, project constraints, project assumptions, initial project organization, initial risks identified, scheduled project milestones, funding limits, cost estimates, project specifications as well as other pertinent information. In addition, the scope statement also provides information on management responsibilities relating to changes or variations. In the event that changes do occur in the future, the scope statement sets out guidance on the level of change control management expected in the project. Procedurally, in the event of a change request occurring, the guidance provided include who to apply for the change request to be approved, what information to be provided, and how the process for approving change request is to be operationalized to meet all project approval requirements.

The direct and indirect effects of a change request on a project must be appreciated. For example, a change in the type of finishing materials used for a building project may lead to a change in the vendor because not all vendors carry all types of finishing materials. Following this change in vendor, the initial project organization should be updated and everyone affected by this change kept informed to avoid miscommunication. When completed, the project scope statement can be a hefty document. This has to be so because the scope statement needs to list out everything that is included in a project so that nothing will be amiss as the work progresses. It is necessary, when available, to provide complete details and full information to all stakeholders to avoid misunderstanding. The provision of a large volume of information must however be properly classified and organized so that all stakeholders know exactly when and where to look for the relevant working details they require to guide them to complete their tasks. In this context, the work breakdown structure is a useful platform for classifying and organizing project information.

h. Work breakdown structure

In an environment with a wide range and variety of information, it is necessary to adopt a classification system for the user to be able to search for and retrieve the information that is needed easily. The library is such an environment where an appropriate classification system that is widely recognized is adopted for the librarian to shelve the books and for the reader to be able to readily find the books needed. The work breakdown structure for a project fundamentally serves the same classification purpose. However, instead of books, the work breakdown structure subdivides a project into components that in turn comprise all activities that make up the components. In essence, the work breakdown structure is a basket of components and activities that create the project.

In the construction industry, the work breakdown can be classified appropriately according to different project types that can either be in the form of building works or civil engineering works. Where relevant, the project manager can make use of past completed projects that are similar as a template to work on the new project. There are also established templates can be adopted for reference for the project manager to build up the work breakdown structure. One example is the Building Cost Information System structure used by the Royal Institution of Chartered Surveyors in the UK. At the first level of analysis, the work breakdown structure contains key components that make up the project. Each component in turn provides the platform for the project manager to identify all the activities that are needed to build that component.

In the construction industry, an example of a key component can be the foundations. The activities that make up the foundations can comprise excavation, piling, pile caps, ground beams and earth backfill. Each of these activities in turn has its own sub-activities that must be identified. For example, in the case of pile caps, the sub-activities can include cutting off excess piles, erection of formwork, earth tamping, laying a screed layer, placing reinforcement bars and casting of concrete. Depending on the complexity of the project, the work breakdown structure can consist of a few layers of information and supporting work details. The amount of details becomes more pronounced as the project manager moves from one level to another level of the work breakdown structure. Classifying project details according to components and activities, at different levels, is necessary as this shows a systematic breakdown of the deliverables to the project manager.

The extent to which the levels are presented depends on the complexity and magnitude of the project. More complex and larger projects therefore require more levels of details for ground operations. In this context, the work package is the smallest element in the work breakdown structure. The work package is given a unique reference number, for example, Work Package 8.5.4.1 so that all relevant stakeholders involved with this particular task are able to readily identify and find the exact element in the work breakdown structure. Depending on the project nature, there are rules of thumb to determine what can make up the work package which is the smallest component in the work breakdown structure. In some smaller projects, a work package is to be completed in one work day of eight hours. In other larger project, a work package is defined to be no larger than 80 h spread over ten days.

There are no hard and fast rules on what makes a good work package. Ultimately, a good work package has a well-defined start and end, and has a work scope that is readily identifiable and manageable given the resource constraints. Apart from providing detailed information on how the project is to be built up from the various components and activities, an effective work breakdown structure should also be expanded to include a glossary of terms to explain new technologies, systems and operations. For example, the work breakdown structure glossary can explain what a clean room, as a component in the project used in the context of a precision production facility is all about. The work breakdown structure also explains what each work package entails, its nature, the resources, estimates and time needed for its completion. Through the work breakdown structure, the relationships between various components as well as work packages are also identified.

This provides an important input to project integration management for stakeholders to appreciate what tasks need to be completed before another task commences. If there are changes to the project scope, the work breakdown structure must also be updated correspondingly to reflect these changes. Updating the work breakdown structure when there are changes in the project scope is critical because the work breakdown provides inputs to identify the activities. Appropriate resources are then planned for and assigned to the activities identified. In so doing, the risks associated with these activities as well as the resources assigned are also identified and evaluated alongside the attendant cost implications. Cost estimates and cost budgets are then derived. Decisions relating to these systematic developments for the projects are possible only with the work breakdown structure.

5.3 Managing Scope Change

a. What if there are changes to the project scope?

While it is acknowledged that the only constant in life is change, it is however not beneficial for projects to suffer from the onslaught of having too much change that impedes progress. This is particularly onerous when changes happen because of poor judgement, fickle-mindedness or ignorance on the part of some project stakeholders. Likewise, frivolous changes can adversely affect project quality, time and cost targets. Nevertheless, where genuine change requests are made, for example, arising from unexpected geological conditions for a building project, such changes must be handled systematically to track, monitor, manage and review how these eventually affect the project scope. Correspondingly, a reliable system is needed to plan, execute, monitor, control and complete the project tasks.

A change control system is therefore necessary to manage change requests as and when these occur. The change control system helps to ensure that scope changes have indeed happened and agreed upon by the relevant project stakeholders. Through an integrated change control system, the stakeholders affected by the change then work together to manage the change as and when it happens. For example, the onset of poor soil conditions in excavation works for a building project may require emergency new works to be undertaken to stabilize the ground conditions. This may require a change in the manner in which the ground is to be excavated and supported quickly to prevent further collapse.

Not all changes occur because of unforeseen circumstances or emergencies. More often than not, the desire or need for change can come directly from the project team members and clients or indirectly from other external stakeholders. In the latter, this can occur when there are changes in the existing regulatory framework that mandates projects to comply with new requirements such as those pertaining to workplace safety and public health. In larger organizations, the project team may want to consider developing an automated decision support system for management to monitor and track changes on a real-time basis.

b. Dealing with scope change

While excessive changes in the project scope should be avoided, it is inevitable that legitimate and reasonable scope change will still occur as the project progresses. When this happens, such change requests should be made with supporting documents to establish the reasons for the change and why it is needed. In addition, going beyond the supporting documentary evidence, it is also necessary for the project manager to determine how such a change request is likely to affect the project scope and its corresponding processes. Following from the change request, additional planning should take place to assess how the change in project scope is likely to affect the schedule and if there is a budget to accommodate the change. In scheduling, the effect of the change is not only direct on the immediate stakeholders. There can also be indirect effects on other peripheral stakeholders whose job scope and related tasks may need to be adjusted to account for the change request. Furthermore, where budgeting is concerned, not all change requests incur additional costs.

In some cases, there may be cost savings for the project arising from the change requests. Where additional funds are needed to account for the approved change requests, the project manager should also at the same time source for additional funds from the client, tap on the contingency sum or transfer funds from other parts of the project. Apart from the impact on project duration and cost, change requests may also give rise to additional risks not previously identified before the requests were made. These risks should be evaluated and their corresponding influence on time and cost noted for mitigation measures as appropriate. A good reference source is the work breakdown structure where individual work packages have earlier been identified. By making reference to the work breakdown structure, the project manager should be able to determine which work packages need to be expanded or trimmed as a result of the change requests.

c. What trigger change request?

There are many reasons why change requests occur that can affect the project scope. Some reasons can be attributed to omissions and oversights on the part of stake-holders as a result of which change requests are made to plug the gap. Availability of more up-to-date market information can also be a cause for change requests made by clients and consultants to use the latest products or technologies in the project. Change requests can occur at any time during a project although it is generally the case that changes are easier to implement at the start of a project than towards the end of a project. After a project has started, regular performance reporting can also trigger change requests. Such performance reports may highlight delays or cost overruns in the project. In such a scenario, change requests may be made to trim the project scope in order to expedite progress or to facilitate cost control. Furthermore, regular performance reports may also point to both corrective and preventive actions taken to improve product correctness or quality levels.

Hence, there is no one standard outcome from change requests that may lead to additional or reduced work. What is however clear is that adjustments need to be made to the project when responding to change requests. Such adjustments are frequently seen as distractions even though the change requests are reasonable and legitimate. The triggers for change requests can come about because of the client's desire to enhance value add through the adoption of the latest technology available. The trigger can also be caused by external events such as the introduction of new regulatory requirements which the project must comply with. Although undesirable, omissions, errors and oversights may also trigger the need for change requests in projects. Finally, change requests may also be triggered by responses to risks identified when the project runs its course.

d. Documenting change requests

Regardless of whether the change requests come about because of external or internal causes, the instructions for such changes can be given either verbally or in writing although the latter is preferred. These would be in response to changes that either impact upon a project directly or indirectly, or are legally binding, mandatory or optional where compliance is concerned. Documenting change requests is therefore an important workflow for the project manager. It is in this context that an appropriate change control system should be implemented to facilitate tracking. For large projects where change requests can be voluminous, the tracking should be automated using computer systems dedicated for this purpose. The computerized change control system should be able to catalogue all such requests and to follow up with the necessary paperwork for all stakeholders affected by the change. The system should enable the project manager to seamlessly track and monitor the progress of all change requests from inception to completion.

Depending on the additional budget likely to be incurred by the change, the system should also be able to route the request to the appropriate authority for approval. The seamless tracking capability of the change control system is crucial for project integration management. This is because a change is likely to affect not only the direct stakeholder but also other peripheral stakeholders both upstream and downstream from where the change request was first initiated. The integrated change control system should therefore inform all relevant stakeholders of the change and how the request if approved is likely to affect their own workflow. If external suppliers are affected by the change, the request should be checked against the contractual provisions agreed earlier with these suppliers. If necessary and depending on the magnitude of the change, the existing contractual provisions can be continued, re-negotiated or terminated as the case may be.

The suppliers should be informed as early as possible of the impending change for them to facilitate their own planning for the project. A change request therefore affects the project management processes relating to initiating, planning, executing, monitoring, controlling as well as closing. All stakeholders, including suppliers, will be affected by change requests. When approved, iterative planning is needed additionally by all affected project stakeholders to cater for such changes. In summary, when approvals have been given for change requests, all affected stakeholders must be notified as part of project integration management. The project scope is then updated via the relevant portions of the work breakdown structure to reflect the new changes. At the same time, the project manager should also reflect on why the changes have occurred. If the occurrence is due to low quality standards, poor performance or unacceptable deliverables of stakeholders, then the corrective actions taken in response should also be documented as lessons learned for future references.

The updating of lessons learned is an important feedback to avoid a repeat of such an occurrence in future projects. More specifically, the project manager should take note of the adverse effects such corrective actions have on scope, cost, time and quality targets and the measures implemented to mitigate or avoid such mistakes in future. Correspondingly, the project baselines are updated to reflect the effects caused by the change requests that arose out of corrective actions. Proper documentation should be kept to avoid misunderstanding just in case there are future disputes relating to the corrective actions taken through approved change requests.

5.4 Verifying Project Scope

a. The need to verify project scope

Project scope management ensures that only the work required for the project is completed, no more and no less, that matches the vision and functions of the project client. This is facilitated through elaborate scope planning that lead to the scope statement which can be a hefty document for large projects. When the project commences, the client needs to be assured regularly that the scope of work is indeed completed as agreed earlier. This assurance is provided to the client by verifying the project scope diligently at every stage of the project. Scope verification is therefore the process through which the client obtains such assurance for accepting the project deliverables.

Scope verification is an on-going process. It happens at pre-defined stages of a project such as at the end of each major phase or when major deliverables are made. In the construction industry, apart from daily site inspections, the workflow leading to monthly certified progress payments is often the primary exercise for scope verification to take place formally. The monthly progress payment exercise as certified by the consultants is meant to ensure that the project deliverables are aligned with the scope agreed with the client. Site inspections and site walk-about are two modes that can be adopted for scope verification. By walking through the building site and inspecting the works that are either completed or in progress, an assessment can be made of the quantity as well as quality of work completed in that month.

Scope verification is therefore more than just quantities alone. It is also a check or control on the quality delivered. If the site inspection shows poor quality standards, the affected work will be rejected and scope verification is deemed to have failed even if the quantities are correct. Scope verification is closely aligned with the project quality plan. Site inspections of the actual works completed compared to the project quality plan allow the project manager to check for correctness as well as completeness of the work agreed with the client. Apart from the project quality plan, other documents should also be referred to by the project manager to clarify outstanding issues identified during the scope verification exercise. These documents can include specifications, drawings, building plans, bills of quantities and architectural models that define what the project scope is. However, site inspection is more than just visual inspection in the physical building site. Inspection can include measuring the actual quantities of claddings installed, examining the workmanship quality of floor finishes, and laboratory testing of concrete cubes.

These are all necessary steps to ensure that all the deliverables meet the client's requirements as set out in the project scope. It is often the case that joint inspections are made together by both the builder and the client or his representative. The benefit of joint inspections is that consensus on acceptable scope verification can be made readily and areas that need to be rectified can be identified at the same time for corrective actions to be taken promptly. Inspections that include site walk-about, audits and product reviews serve to deliver this benefit. At each stage of the scope verification exercise, the client formally accepts the deliverables once the scope has been verified for completeness and correctness. Payment for that portion of work is then made, leading to closing. In the event that no formal acceptance is forthcoming, scope verification may be conditional depending on the outcomes of the specific work concerned.

If corrective action is required, scope verification is repeated when rectification is completed satisfactorily. Otherwise, that portion of work is deemed to have failed, leading to cancellation of the project in extreme cases. Taking the middle ground, however, that portion of work whose scope verification is not formally accepted may be put on hold pending further decisions by the client. In some cases, even though scope verification indicates fulfilment of the entire product description, it is also possible for a client to be unhappy. Nevertheless, so long as there is completeness and correctness, the project is considered to be technically complete even if the client is unhappy.

Proper documentation in the form of a written agreement is recommended to prevent such a situation from happening that can be detrimental to the builder or supplier. Finally, scope verification is also important in the construction industry from the regulatory perspective for fire protection and to safeguard public health and safety. Before a building can be occupied, a temporary occupation permit needs to be obtained from the relevant building authorities. The temporary occupation permit is issued only after the authorities are satisfied that the building has been completed according to the building plans that have earlier been approved. This is facilitated through a joint site inspection between the authorities and the client or his representatives. The joint site inspection leading to the successful issue of the temporary occupation permit forms part of project scope verification in the construction industry.

5.5 Revision Questions

1. What is the role of the project charter in scope management?
2. How is the project scope established?
3. To what extent can project scope be developed based on past completed projects?
4. What should the project manager do if a project scope for an entirely new project without any past references is to be developed?
5. What are the consequences for not adhering to the project scope (i.e. unauthorized additions or omissions)?
6. Why is the project scope statement a point of reference for stakeholders?
7. Is there a difference between product scope and project scope?
8. What constitutes progressive elaboration for project scope planning?
9. What is a project scope management plan?
10. Why are changes and variations better managed with a scope management plan?
11. What does a project scope management plan do?
12. Is the work breakdown structure part of scope definition?
13. What is the contribution of the work breakdown structure to scope definition?
14. What are some of the inputs for scope definition?
15. Why is product analysis beneficial for a project?
16. What are some of the activities needed to create the scope definition?
17. Why is it necessary to examine the scope statement?
18. Why is the project scope statement often a hefty document?

19. How does the work breakdown structure contribute to development of the project scope statement?
20. Is it true that the work breakdown structure differs from one building project to another building project?
21. Why is it necessary to verify the project scope as the project progresses?
22. What are some of the methods available to verify the project scope as the project progresses?
23. What happen after the project scope has been verified successfully?
24. Why is it beneficial to protect the scope from change?
25. What is the role of the work breakdown structure in project scope change?
26. Should all change requests be approved?
27. What are some of the legitimate reasons for changes that are needed in project scope?
28. What happen after change has occurred?
29. What are the follow-up actions prompted by a change control system?
30. In project scope management, why is it necessary for scope planning to come before scope statement?
31. Why are correctness and completeness two important considerations for scope verification?
32. When can scope change take place without seeking prior approval from the client?

Chapter 6
Project Time Management

6.1 Introduction

a. Time management is about planning

Where time is concerned, project management has been described as a nine-to-five job. This is so because project management comprises of the nine knowledge areas that are underpinned by the five processes. Planning is however the key driver behind good time management. It has often been said that if a person fails to plan, that person plans to fail. Time management is about understanding the type, nature, magnitude and intensities of the activities involved in a project and scheduling these activities to meet the various deadlines. To do this, the project manager therefore needs to possess good understanding of the activities in the industry concerned. The activities in the banking industry are clearly different from the activities in the manufacturing industry.

Even in the manufacturing industry itself, activities are also different depending on what goods are being produced. These can range from the production of automobiles to the production of claddings used in the construction industry. Hence, the project manager needs to possess not only a good understanding of project management practices but also competent domain knowledge of his chosen industry. It is generally acknowledged that the construction industry is by far one of the most complex industries in a country. This is especially so for mega complex building projects or cross-country civil engineering road construction. Each site location is different even if the building type is the same.

The transient nature of the construction industry is such that stakeholders come together for a short period of time and disband when the project is completed. There are also many different construction methods across a wide range of building elements that span from substructure works, superstructure works, architectural works, mechanical and electrical works, as well as external landscaping works. The project manager should therefore possess competent domain knowledge in construction processes and practices before he is even able to start planning. Take the

© Springer Nature Singapore Pte Ltd. 2018
L.S. Pheng, *Project Management for the Built Environment*,
Management in the Built Environment,
https://doi.org/10.1007/978-981-10-6992-5_6

example of foundation works where there are many choices available depending on the ground conditions and building types. In excavating for foundation works, there are also choices between different types and capacities of excavation plant and equipment. In the planning process, decisions need to be made relating to the number of such plant and equipment to be used and for how long.

There is also the need to strike a balance between costs and time when planning for the use of such excavation plant and equipment. Issues relating to purchasing, leasing and renting, need to be thought through. In some cases, such decisions can be made more easily because such excavation plant and equipment are readily available within the organization. Such decisions also incur corresponding risks which should likewise be considered. Multiply this by the large number of different elements in a typical building and it is not difficult to appreciate the extent of planning work that the project manager has to undertake for realistic time management. Take the example of an integrated resort complex. A typical mega resort complex can include supporting infrastructures, utilities, hotels, shopping malls, restaurants, recreational facilities, car and coach parking. To facilitate planning for time management, these should be divided into phases to render the process more manageable for the project manager.

The number of activities by now would have skyrocketed to their thousands which render manual means of planning untenable for the human mind to comprehend. In such cases, project management software programs should instead be adopted for planning and time management. It is also opportune to consider hiring a planning manager, planner or scheduler to work on these tasks on a full-time basis. In a nutshell, time management is, among other things, about planning ahead. It relies on numerous inputs to plan, monitor and control the project schedule.

6.2 Definitions of Project Activities

a. Need to define project activities

All projects are made up of activities. Activities are identified from major project components such as foundations and plumbing systems in a typical building project. Components are in turn identified and scoped using the work breakdown structure. Projects are temporary in nature; so are the major components and their corresponding activities. It is through planning and time management that the start times and end times of these activities are determined. Before a project commences, all activities that are needed for the successful completion of the project must first be identified.

Project time management therefore starts with defining and identifying these activities. The activities thus identified should then all be sequenced in a logical manner that is technically sound. For example, wall papering can only commence after the drywall partitions have been installed. The time duration needed for completing each activity should be estimated. The duration can of course vary depending on the amount of resources that are used for that activity. If more manpower is used, then the installation of the drywall partitions can be completed earlier and vice versa. All the activities together with their estimated durations should not be planned to take place sequentially. Doing so will stretch the overall project duration. It is often the case that some activities can be carried out concurrently at the same time. In the case of building services, installation of ducting for the air-conditioning system can take place at the same time as the installation of electrical cables and light fittings at the ceiling level.

Based on both sequential and concurrent planning, the overall project schedule is then developed which shows how long the project is likely to take to complete. The project schedule also shows the start times and end times of all the different activities and how these activities relate to one another. This facilitates schedule control for the project manager to monitor and control work progress.

b. Paying attention to details

Planning is by far the most important process for project time management. Once the plans are set up, the project manager uses these plans to execute, monitor and control the activities. When more details are given to planning the activities, there will be lesser opportunities for mistakes and errors. Nevertheless, it should be recognized that defining and identifying the activities are not a stand-alone task. This can be influenced by inputs from and choices in the other eight project management knowledge areas. Activity planning for time management is therefore influenced by considerations relating to scope, cost, quality, human resources, risks, communications, procurement, and integration management. For example, in the case of procurement management, there is often a choice to either purchase from an external supplier or make the product in-house depending on the benefits associated with each of these two choices.

The activities for this product therefore depend on the choice made. Having identified the activities the planning process should progress to consider the sequence of these activities. Sequencing these activities can also be influenced by both internal and external events that are not within the control of the project manager during the planning process. For example, the manpower information from the human resource department is internally unavailable for the project manager during the planning process. Externally, the procurement of products from overseas relies on the timely information provided by their suppliers to the project manager. Most of these events can also be classified in the known or unknown category. In the former, the project manager can rely on historical records in the organization for similar projects in the planning process.

Unknown events such as labour strikes, political unrests in a foreign country where the products are procured, as well as inclement weather should also be factored in by the project manager for schedule development. In situations where there is no prior experience to fall back on, the project manager needs to assess the probabilities of such unknown events unfolding based on the most optimistic and the most pessimistic scenarios and their respective effect on the time-line.

c. Preparing the activity list

All activities identified for planning must be relevant for the project scope. Identification of activities should therefore start with the project scope statement which sets out only the work required to meet the client's requirements, no more and no less. The activity list is best organized using the work breakdown structure which provides a systematic presentation of the components needed to deliver the project. For large complex projects, it is useful to develop and implement a dedicated real-time project management information system that can be shared by all stakeholders for planning purposes. Other readily available scheduling software can also be adopted if there are no specific requirements that need to be tailor made specially for the project team. There is also much mileage to be gained by referring to past records of projects completed in the organization.

Through these documents, the project manager should be able to understand how things are done in certain ways based on organizational policies and preferences. Such historical records also provide useful templates and checklists for the project manager to identify the activities needed to build up the various components that make up the work breakdown structure. Where necessary, the project may need to be planned in phases for activities to be better defined, scheduled and developed as part of the overall project management plan. Relying on existing templates is a useful start for planning activities. There is no point in re-inventing the wheel if similar projects have already been completed in the past by the project team. The existing work breakdown structure, components and activities can be adopted to serve as templates for the new project. Tweaking and modifying some aspects of the existing templates are of course essential to account for the differences between the current project and the past completed projects.

d. What the planning templates include

For project time management, existing and newly developed templates for planning purposes should include the following elements. The templates should show the actions that are needed to complete the project scope, including considerations relating to how scope change should be managed and how scope verification is to be carried out as the project progresses. The templates should set out the methods used to achieve the various project deliverables. For example, in the construction industry, consideration is to be given to whether cast in situ concrete, precast concrete or structural steel construction is to be adopted for the superstructure of the new skyscraper. Once decided, the corresponding resources required for the chosen method of construction should be reflected in the template.

The resources for the chosen method should include details relating to materials, machineries and manpower needs. Based on past experiences, the duration needed to complete the activities identified is then determined. Depending on the activity type, duration can be expressed in hours or days. The project manager can rely on completed time study and motion study to establish the labour constants used to estimate the duration needed to complete an activity. Labour constants are known outputs of work based on a definitive set of resources such as a team of three painters or a team of five bricklayers. Labour constants in these two cases can be expressed in square meter per hour or square meter per day. The output constants for other resources such as excavators can also be retrieved from past records. In this case, the output of an excavator can be expressed as cubic meter per hour for the volume of earth excavated.

While the labour constants and output constants can be retrieved from past completed similar projects, it should however be noted that not all projects are the same. The risks associated with past completed projects may not be the same as the current project because of site location and client type. For this reason, planning for time duration should also take into account the risks and how these may possibly impact on timely completion of the activity. Activities that are grouped together for a specific project deliverable (for example fencing) can be placed within an identifiable work package with appropriate descriptions and supporting details. Such provisions would allow the project manager to mobilize all resources to complete these activities in the near future.

e. Dealing with future planning

Work on the activity list can start after the identification of major project components and the creation of the work breakdown structure. For typical building projects, there can be many different sets of activity list that must be compiled to be used as the fundamental tool to derive the project schedule. The activity list is therefore an extension of the components and the work breakdown structure. However, for large complex projects stretching over several years, it may not be feasible to plan for work in great details that happen only in the distant future. For this reason, a judgement is to be made on what needs to be planned to the greatest details for works that are imminent.

For works that occur in the distant future, planning is still required but parked at a high level for actions to be taken only closer to the date when these works are needed. This makes sense because details for such works in the distant future may not be readily available currently, or even if available are likely to be changed closer to the date of their implementation. Planning for works in the distant future should therefore take on the progressive elaboration process. In this case, operational details are progressively sourced and elaborated as the project progresses. This is often known as rolling wave planning wherein planning first focuses on the more pressing and imminent activities as the project progresses. At the same time, an activity that is in the distant future should also be planned but with a "to be decided later" note attached to it.

Closer to the day when that specific activity requires more details for implementation, an alert encapsulated in the planning software program should be sent to the project manager to take further timely actions. Such details may relate to for example the wall paper patterns, colours of the floor tiles and light fixture designs. It is necessary at that stage to consult the client for his choice to be selected from the different patterns, colours and designs then available. The planning software program should be able to build in a management control point to trigger an alert to the project manager in the future when additional planning for an activity is around the corner. The trigger can be in the form of a visual alert such as a flag or an email message sent to the project manager. Once the project manager receives the alert, further detailed planning for that activity can then take place immediately.

f. Systematic planning using the work breakdown structure

For planning to take place effectively, it must be undertaken systematically and comprehensively. The work breakdown structure provides an effective platform for planning. Based on the scope statement and past completed similar projects, all the components required for the current project are then identified by the project manager. This is essentially a thinking process. An example of a major component in a typical building project is the plumbing system. The activities that are needed to install the plumbing system to full operational condition are then identified. Examples of such activities are the laying out of pipework within the building, installation of an external water meter and connection to the water main following approval from the relevant authorities.

The activity list comprises all activities needed to produce a component. In the activity list, the nature, type and characteristics of each activity should also be identified so that appropriate resources can be allocated for its completion. For example, excavation can be done manually or by using an excavator. If it is to be done manually, the activity list for excavation should include the number of workers and the tools required. These supporting details are important not only for planning and time management but also capture the background reasons why a specific construction method and resources were used for that activity. The supporting details associated with an activity can include the assumptions made at the time of planning when no information is readily available. The details should also highlight the constraints faced. For example, one constraint leading to the use of a silent pile driver is to avoid excessive disturbance to residents in a built-up neighborhood.

Another constraint may relate to limited working hours for a school refurbishment project during the examination period. The supporting details should also provide the reasons for a specific work package that may for example involve bringing in skilled artisans from Hunan, China to recreate intricate designs for an old Chinese temple in Singapore. In addition, the supporting details should highlight information that is pertinent to practices in the construction industry. This may relate to specific Singapore Standards for concreting works or Codes of Practice for electrical installation works in the current project. It should however be appreciated that planning for a large complex project is not without its challenges. There will be situations where information is not readily available at a time when it is needed or worst still the wrong information has been tendered. Careful checks of the activity lists for errors, discrepancies or inadequacies should therefore be made for correctness as the project progresses. Where such errors, discrepancies or inadequacies have been detected, these need to be corrected and the work breakdown structure updated immediately

6.3   How Activities Are Mapped

a. Developing the schedule

The activity lists identify all activities that are necessary for completing the project. Having identified the activities, the next step is to arrange them in a manner that flows logically. Not all activities need to be arranged in a sequential manner. Doing so will unnecessarily increase the duration for project completion. Some activities can be arranged to run concurrently. Whether activities are to run sequentially or concurrently depend to a large extent on their relationships, the constraints faced or the preference of the project manager.

The flow of activities can be planned manually for small projects. However, the manual process is no longer viable for large complex projects. In this case, appropriate planning software programs should be adopted which can also help to readily facilitate future revisions. A combination of both the manual process and computer-driven process is often used where bite-size list of activities are first arranged manually using pen and paper or sticker notes. When completed, the sequence of activities with arrows showing their relationships is then inputted into the planning software program.

b. How to start?

The starting point for activities planning rests with the project scope statement or the product description from where the work breakdown structure is then derived with the corresponding components identified. Each component in turn identifies all the activities necessary for the completion of that component. Once these activities have been identified, sequencing of these activities takes place. There are however three dependency considerations that can influence the sequencing of activities. Compulsory dependencies anchored on hard logic do not offer any choice for planning as these are often regulatory or technical in nature. For example, from a regulatory perspective, demolition works cannot commence unless and until a permit has been obtained from the building authorities. Similarly, from a technical perspective, concreting cannot take place until all formwork and reinforcement bars are in place. Likewise, interior decoration works cannot start until the roof is completed.

On the other hand, discretionary dependencies anchored on soft logic do offer some choices for activity planning. The sequencing is discretionary due to preference or unique project conditions. Here, the project manager can decide if connection of services can take place at the same time when doors and windows are being installed. Or he can decide to connect the services only after the doors and windows have been installed alongside with the erection of external fencing. Thirdly, external dependencies may or may not offer any choice for planning. This depends on the nature of the external dependencies which is not within the control of the project manager such as rain, flooding, delivery of equipment from a supplier or the impending introduction of a new building regulation. However, if mitigation measures have been taken, the project manager can exercise some degree of control over external dependencies. For example, construction firms have been known to erect big canopies over the entire building site to mitigate the adverse effects from inclement weather.

Collectively, the compulsory, discretionary and external dependencies do affect the manner in which activity sequencing is carried out. In addition, these dependencies are also tied to specific milestones which the client sets for the project team. For example, such milestones are set out in the progressive payment scheme for home buyers of new condominium apartments. Under the scheme, the home buyer needs to make progress payments to the developer-client following the completion of each milestone in the new building project. When foundation works are completed, the home buyer pays 10% of the purchase price to the developer-client. Likewise, when the reinforced concrete frame is completed, another 10% of the purchase price is to be paid by the home buyer to the developer-client. The developer-client is more likely than not to hold the construction firm to the various milestone time-lines in order to facilitate cash flow. Such milestones can therefore influence how activities are being sequenced to avoid unnecessary delays.

6.4   The Network Diagram

a. Showing relationships between activities

After all activities have been identified, it is necessary to establish their attributes with supporting details. These attributes describe who is responsible for the completion of the activity and what comes before and after this activity. In addition to showing the successor and predecessor, the supporting information should also present the lead time and lag time associated with the activity relative to its successor and predecessor. Thereafter, the activities are sequenced to show their relationships from start to finish of the project. The outcome of this exercise is the network diagram which uses arrows to show the relationships between activities and how long it takes to complete each of these activities as well as the entire project.

In this case, the activity is indicated on the arrow itself, connected by nodes which show the dates. However, there can be instances where a node is linked to two succeeding nodes with two arrows to show their relationships. In order to show more than two relationships, there can be an activity on one arrow with another arrow without an activity. The arrow without an activity is known as a "dummy" activity with no time duration. Dummies are simply used in the arrow diagram method to show the logical relationship between two activities without invoking time duration. Dummy activities are indicated as dotted lines while real activities are indicated using normal lines in the arrow diagram method.

The network diagram also shows the effect of delay one activity has on another succeeding activity and on the overall project duration. In the process of creating the network diagram, missing activities may also be surfaced which were not identified earlier in the work breakdown structure. Updating the activity lists should therefore take place simultaneously to account for the missing activities. However, the network diagram method using nodes and arrows is only able to show the "finish-to-start" relationships between activities. Hence, this is a major weakness of this method. In real life, relationships between activities can be more complicated than the simple but most common "finish-to-start" relationship. For this reason, other planning tools such as the precedence diagram have been developed to allow for other complex relationships between activities to be shown.

b. Annotating specific relationships

The network diagram simply shows the starting and ending relationships of activities. However, the reality is that there can be more complex relationships between activities beyond mere starting and finishing points. This is a limitation of the network diagram. To overcome this limitation, the network diagram has given way to other planning tools such as the precedence diagram where more complex relationships between activities can be accounted for. There are four such possible relationships to consider when planning for activities to take place. The first is the most common "finish-to-start" relationship. Examples of this relationship include the completion of brick wall before plastering can start and completion of the wiring system before installation of light fittings can commence. The second is the "start-to-start" relationship between activities.

Hence, in an external landscaping example, painting to wooden fencing can start once the erection of wooden fencing commences and some of the fence erection has been completed but not all. Similarly in renovation works, priming can start once scraping off the old paint commences. In this relationship, an activity can start once another related activity has commenced. The third is the "finish-to-finish" relationship between activities. An example is the testing of the plumbing system. Once the installation of the plumbing system has been completed, testing of the system can start and finish at about the same time by turning on the water supply.

Another example is the installation of a new automated card access system for an office over the weekend. The installation as well as testing of the system must both be completed by Sunday evening in time for use on Monday morning. The fourth is the "start-to-finish" relationship where related activities are completed just in time to meet a stipulated deadline. An example relates to the application of a temporary occupation permit from the authorities for a newly completed building. The application for the permit involves a joint site inspection from a building officer from the authorities together with the builder. The builder is not likely to wait until all outstanding defects have been rectified before applying for the permit. The application for the joint site inspection can start while defect rectification work is still on-going. However, the builder needs to ensure that all rectification work is finished before the joint site inspection takes place.

c. Lead and lag times

While the norm in creating the network diagram is anchored primarily on the "finish-to-start" relationship, this relationship can be subject to lead and lag time effects as well. Lead time means a downstream or succeeding activity can begin before the upstream or preceding activity is scheduled to end. For example, preparations for laying floor finishes for a new house can start two days before window installation is scheduled to complete. Conventionally, internal floor finishes are laid only after the completion of window installation. This is to ensure that the newly laid floor finishes will not be damaged by driving rain into the building. If preparations for laying floor finishes commence two days before the scheduled completion of window installation, there is a saving of two days from the project duration. Time is deducted from the downstream floor finishes activity. Lead time is therefore a negative value to the project duration. However, having too many activities operating on this negative mode can lead to increasing risks to achieve more lead time.

The risks can increase because the project is trying to achieve too many things in too short a time frame. The site can become more congested with too many people doing too many things at the same time. This may cause accidents to happen. On the other hand, the "finish-to-start" relationship between activities can also be subject to the effects of lag time which has a positive value and add time to the project schedule. Lag time is typically waiting time which cannot be avoided because of the nature of both the upstream and downstream activities. For example, painting cannot start immediately after plastering is complete. There is a lag or waiting time for the plastering to dry and harden before painting can start. Similarly, after the waterproofing screed layer has been laid in the bathroom, there is waiting time for the screed to dry before the ponding test takes place to ensure water-tightness. After the 24-hour ponding test is completed successfully, floor tiles can then be laid in the bathroom. Lag times such as these can therefore increase the project schedule. In time management and planning, it should be remembered that lead removes time and lag adds time to the project duration.

6.5   Time and Resource Requirements

a. Effects of resources on time

Projects require a wide array of resources for their completion. When the work breakdown structure identifies the components and activities needed for the project, the resources required to complete these activities are also identified. Examples of resources can include materials, plant, equipment, tools, tradesmen, design consultancy service, software and staff training. More often than not, there are choices in selecting resources. For example, concrete can be mixed on site by the builder, purchased from a ready-mixed concrete supplier or delivered to site as precast concrete components for installation. There is also a choice for the builder to either "make" or "buy" the materials with consequential influence on procurement management and the procedures needed to acquire these resources.

Resources do influence project schedule depending on the selection made. A choice between manual excavating using labour and mechanical excavation has implications for time and costs. Selecting mechanical excavation leads to faster completion but possibly increased costs. While resource type can influence time, the manner in which such resources are brought into the site can also affect time. It does not mean that introducing more resources into the site will result in time saving from the lead effect. On the contrary, introducing more resources into the site indiscriminately can result in site congestion that can in turn adversely affect productivity. Logically, if two bricklayers can complete a wall in a day, then by linear extrapolation, having four bricklayers can mean that the wall can be completed in half a day.

This is however not always true all the time. When there are too many people in the site, this can potentially lead to saturation where it becomes too crowded for the bricklayer to function productively. This can then be counterproductive with diminishing returns as more resources are added to an already saturated situation. There can however be situations where the deadline for the completion of a project is dictated by the client that cannot be changed. This can happen when the client hopes that the completion of a new condominium project is timed to catch the anticipated upswing in the property market. In such a situation, the project manager needs to work backward from the stipulated deadline to identify the activities and the resources required to complete the project successfully.

b. Activity durations

The durations for all activities must be estimated before the project schedule can be completed. The activities for a project are identified from the work breakdown structure. These activities are then sequenced in a logical manner that makes construction and installation possible. The sequencing is however not able to show the overall project duration unless and until the time needed for completing each activity is known. To estimate the durations needed for these activities, decisions relating to the resources to be used have to be made. This is an important step because the type and nature of resources used for an activity determines in a large way the estimated duration needed to complete that activity. For example, timber formwork while cheaper takes a longer time to erect. On the other hand, system formwork can be erected more quickly but is likely to cost more than timber formwork. Once the type of resources needed for an activity have been decided, the duration for that activity can then be estimated. This applies to all activities with their corresponding resources identified and their durations estimated.

Collectively, all these activities are then sequenced together to complete the project schedule which then indicates the estimated overall project duration. It is clear that the duration of an activity is determined and estimated from the resources needed and used for that activity. Domain knowledge of the field or industry is therefore an important attribute of the project manager. Without this domain knowledge, the project manager would be unable to identify what resources and in what quantities are best suited for each and every activity. In unique situations where the project manager is not familiar with the technical know-how, expert judgement provided by other professionals is needed to provide the inputs for these activities. Estimates for activity durations are best obtained from the people who carry out the activities. Hence, estimating the duration needed to complete a brick wall is best undertaken by asking the bricklayers for their inputs.

c. Inputs to estimate time durations

The inputs to estimate time durations start with the list of activities that is needed for the project. These activities are in turn identified from the work breakdown structure that shows the various components from which these activities are derived from. Identifying the various components and their corresponding activities from the work breakdown structure is essentially a thinking process. The resources required for completing these activities are first identified. Most resource types provide choices. For example, excavation can be undertaken manually or mechanically. The choices selected have implications for time and costs. The activities required for building a residential dwelling can include among others the architect, engineer, surveyor, bricklayers, plumbers, electricians, painters, bricks, concrete, roof tiles, floor finishes, mobile crane, hand tools, etc.

It is necessary to identify all the resource requirements for these activities in as comprehensive a manner as possible so that nothing is amiss. Each of these resources also has its own characteristics that can influence the activity attributes. It is logical to deduce that the intensity of efforts determines the duration for completion of the activity. Hence, using more resources leading to greater efforts can generally lead to shorter duration for completion of the activity. But this is true only to some extent because more resources converging on a project site at the same time can lead to congestion that hinders productivity. It is also true that some activities have fixed duration for their completion regardless of the number of workers present.

For example, setting up a concrete pump requires a fixed number of workers for its completion. Having more than the required number of workers for this purpose, however, is not likely to hasten the process of setting up the concrete pump. The capacities of certain resources are also fixed. For example, a mobile concrete pump can only deliver a fixed quantity of concrete in an hour which is indicated in cubic meter per hour. With this fixed capacity in mind, the number of ready-mixed concrete trucks sent to the site should be synchronized with the pump capacity. Sending more trucks of ready-mixed concrete to the site in a short period of time is not likely to reduce the duration needed to complete the pumping operation. The synchronization can be facilitated through findings obtained from work study or motion study.

Resources should also be considered for both their capacities and capabilities. For example, a more experienced bricklayer is likely to work faster than a less experienced bricklayer. The use of precast concrete components is likely to lead to faster completion of the building project compared to the use of cast in situ concrete. Much of such information relating to the types of resources appropriate for the project, their capacities and capabilities as well as the time estimates is readily available from existing records. These records are kept within the organization such as the project files of other similar projects completed recently. Such in-house records are generally more reliable. Records can also be obtained from external sources such as those available from commercially available estimating databases. Such databases show the different types of resources and their corresponding time estimates. This can show for example how many bricks a team of three bricklayers can complete laying in one hour.

Where such recorded information is not available, it is useful to consult other project team members for their inputs based on similar projects that they have undertaken in the past. It is also necessary to consider if the methods and the resources chosen for the activity post any risks during execution. If there are, risks mitigation measures have to be taken. These measures in turn can influence the estimated duration needed to complete the activity. For example, excavation for a building project near a river should include the need to pump out ground water. Collectively, information relating to all the resources, their associated risks, mitigation measures and their time estimates are included in the project management plan for implementation.

d. Expert judgement and reserve time

Some projects may be undertaken by the project manager for the first time for which there are no existing records within the project organization that he can refer to estimate time duration. In such a situation, evaluation by an expert in that field should be sought for his expert judgement to provide a time estimate. An expert can be anyone with the knowledge in that domain area. For example, time estimates can be obtained from a geotechnical engineer relating to difficult substructure construction works. The expertise needed for the construction of a dirt bike racing track can be sourced from someone based in Australia who possesses the knowledge on how this is to be done. Experts are therefore subject matter experts as well as other project team members both within and outside the organization who are familiar with the new activities. While due care and diligence needs to be exercised when estimating time durations, reserve time should also be factored into account for interruptions.

At the personal level, such interruptions can occur because of long meetings, answering telephone calls and dealing with email messages. At the site operational level, such disruptions can occur due to inclement weather or sudden introduction of new regulatory requirements that affect the project. In reality, the project manager should also factor in human behaviour when estimating time durations. In this context, Parkinson's Law states that work expands to fill the time available for its completion. There are hidden times which workers may exploit to take longer than necessary to complete an activity at a leisurely pace. The project manager should also watch out for workers who do not immediately inform management that an activity has been completed. Instead, such workers would keep silent on early completion of the activity and report completion only just before the due date is up. Time estimates can also be affected by procrastination of stakeholders that may trigger a snowball effect on other activities, leading to delays. In responding to multiple demands, workers may resort to multi-tasking which can affect their productivity as their attention span can be adversely affected as they move from one project to another in rapid succession. They may also be stressed out and become tired. All these have the net effect of prolonging the time estimated to complete an activity.

e. Assessing time estimates

Most if not all time estimates are made based on the information available at a point in time. As the name suggests, this is only an estimate of the time needed to complete an activity. Hence, time estimates may not be entirely accurate. These should factor in time variances depending on the assumptions made and the constraints faced. For example, a time variance of plus or minus 2 days for painting means that a 8-day painting work activity can take up to 10 day or 6 days. In evaluating the time estimates for variances, the basis for the estimates and the underlying assumptions should be noted and documented for future reference. The method chosen for the activity should be identified.

For example, the choice of manual excavation instead of mechanical excavation should be justified and recorded. Likewise, the number of workers expected to be mobilized for the execution of an activity such as the number of painters for a painting job should be documented. In evaluating the list of activities in relation to their respective time estimates, the project manager may also discover that some activities are missing. This should be rectified immediately and included in the list of activities. For example, in laying floor finishes for bathroom areas, the activities only listed waterproofing, laying of screed and floor tiles. From evaluating the activities, the ponding test was found to be missing and should therefore be included in the said list of activities.

6.6 Project Schedule

a. Developing the schedule

The tasks completed thus far relate only to the identification of the work breakdown structure, components and activities. The time estimates for all activities have also been completed. It is now time to link all activities together in the form of a project schedule to determine how long the entire project will take to complete. This exercise is unlikely to be a one-off exercise because of the iterations involved. Iterations of the project schedule are necessary because of updates to the lists of activities and the realization of the underlying assumptions made and the constraints faced. For large projects, the iterative nature of project scheduling can become complex. Hence, in such cases, it is imperative that scheduling software programs be used to deal with the iterations and complexities. Apart from the assumptions made in project scheduling, a major constraint relates to the deadlines set by the clients or stakeholders downstream in the supply chain. These pre-determined time constraints have to be factored into the project schedule.

b. Time constraints

There are fundamentally four time constraints that affect the project schedule and ultimately the project duration. The first constraint is the "Start No Earlier Than" requirement. For example, a renovation contractor refurbishing a hotel may be instructed to start noisy work no earlier than 10 am each day. This is to ensure that hotel guests are not disturbed too early in the morning. The second constraint is the "Start No Later Than" requirement. For example, if Independence Day falls on 4th July, then a project for celebrating this occasion should start no later than 4th July. The third constraint is the "Finish No Later Than" requirement. For example, if arrangements for the installation of kitchen equipment have been made for 10th October, then the floor tiles for the kitchen must finish no later than 9th October. This is because without the completion of the floor tiles, installation of the kitchen equipment cannot take place. The fourth constraint is the "Finish No Earlier Than" requirement.

For example, a show-flat that is to be opened for public viewing in a new condominium project must first obtain a temporary occupation permit from the building authorities. If no permit is issued, then members of the public cannot view the show-flat. Hence, if the permit is expected to be issued on 11th April, then preparation works for the show-flat need not finish earlier than 11th April. Opening of the show-flat should not be earlier than 11th April.

c. Identifying the critical path

The project schedule can be developed using the precedence diagram. The Early Start Date and the Late Start Date for each activity can be factored in the precedence diagram. Likewise, the Early Finish Date and the Late Finish Date for each activity are also included using the critical path method. Reference is made to early and late start dates because many activities do have the choice of starting either on Monday or anytime of the week. Consequently, the choice correspondingly lead to the early and late finish dates. By starting early, an activity also finishes early and vice versa. By adopting a "forward" pass and a "backward" pass, the critical path method is able to reveal which activities are considered critical in the first instance.

The common path in which all the critical activities take place is known as the critical path. The critical path is the path with the longest duration needed to complete the project. All activities along the critical path must not be delayed. If such critical activities are delayed, this will cause the project completion date to be correspondingly delayed. The critical path may also shift when there are iterations to the project schedule in responding to the assumptions made or the constraints faced by an activity. Activities not on the critical path are known as non-critical activities. Such non-critical activities possess buffer or slack time that allows some delay in the non-critical activities without delaying the project completion date.

Buffer time is also commonly referred to as float. A Free Float is the total time that a single activity can be delayed without delaying the early start of any succeeding activities. The Total Float is the total time that a single activity can be delayed without delaying the project completion date. Finally, the Project Slack is the total time that a project can be delayed without surpassing the completion date expected by the client. The critical path method is not the only tool that can be used for time management. Depending on the nature of the projects, other tools such as the bar chart and the line-of-balance diagram can also be used. The bar chart is a useful scheduling tool for simple projects or to serve as an overall control chart to provide a bird-eye view of key milestones of the project. The line-of-balance diagram is a useful tool for projects with repetitive activities such as those for civil engineering works in pipe installation or cable-laying over long distances.

d. Compressing the duration

The completed schedule shows how long the project should take for completion. There are however cases where the estimated project duration is considered to be too long by the client. This can happen for example when there is an anticipated upswing in the residential property market and the developer client wishes to complete the new condominium project quickly in order to ride on the upswing to achieve more sales. The project duration therefore needs to be shortened or compressed. This is achieved by "crashing" where more resources are added to activities on the critical path to hasten their completion and thereby reducing the project duration. Crashing the project duration does not however work for activities that have a fixed time frame for their completion. For example, the setting time for cast in situ concrete in the building site cannot be shortened.

Similarly, the time taken to complete the ponding test to ensure the effectiveness of waterproofing in wet areas of a building requires a set duration to complete. There are however other activities whose duration can be shortened. For example, bricklaying can be completed earlier if additional bricklayers are employed for the task. Similarly, the time taken to erect and dismantle formwork for concrete casting can also be reduced if system formwork is used instead of timber formwork. A balance would however need to be made between the bonus the builder is likely to achieve for earlier completion and the additional costs that he needs to incur for crashing the activity duration. Crashing can therefore lead to increase in costs. Another approach to reduce the overall project duration is to fast-track activities. In fast-tracking, activities that are normally undertaken in a sequential manner are arranged to be carried out in parallel or with some overlaps between them. For example, painting is normally carried out only after the floor tiles have been laid. However, in fast-tracking, painting and tiling works can start together provided this arrangement does not lead to site congestion and safety issues. Fast-tracking therefore to some extent increases project risks.

Time estimates also incorporate a range of variances which can be described as the "most optimistic time", the "most pessimistic time" and the "most likely time". Using the Program Evaluation and Review Technique, such variances based on statistical probabilities can be included to correspondingly estimate the "optimistic completion date", "pessimistic completion date" and "most likely completion date". Consideration should also be given to resource levelling and smoothing. This takes into account the amount of resources to be mobilized in order to crash or fast-track activities. It is not desirable to introduce too many resources in a project only to remove these after a short period of time. Such peaks and troughs in resource utilization can be counterproductive. Resource levelling and smoothing therefore make a conscious attempt to even out resource utilization to achieve a steady pace of work in the project.

6.7 Schedule Control and Update

a. Project management processes

Project time management is underpinned by the project management processes that include initiating, planning, execution, monitoring, controlling and closing. After the project schedule has been developed, it needs to be monitored and controlled during implementation. This is because scheduling is an iterative process that takes into account changes that can happen throughout the entire project duration. Changes caused by internal and external events would require updating through a dedicated schedule control system which records the conditions, rationale, effects on time, costs and risks arising from making the change. The system should also be linked with integrated change control for the project so that all stakeholders affected by the change are kept informed. While activities that fall on the critical path merit close attention to ensure these are completed in a timely manner to meet the project completion date, activities on the non-critical paths should also be monitored closely.

This is because activities on the non-critical paths may be delayed but only to the extent that they do not delay the completion of activities on the critical path. If there are significant delays in the non-critical activities, these may in turn evolve to become critical activities, thus replacing the previously critical activities in the original critical path. A new critical path is then created. For large complex projects, the project team should continuously exercise prudence to track both critical and non-critical activities. Where necessary, intervention should be made to apply timely corrective actions to align the schedule back with the original dates to ensure timely project completion.

6.8 Revision Questions

1. Is project time management the same as scheduling?
2. Why is it necessary to identify the project activities in time management?
3. Is relying on an existing template or historical document the easiest way to identify project activities?
4. How are project activities and their respective durations identified for an entirely new project for which no historical records exist?

5. Why is "rolling wave planning" part of progressive elaboration in project time management?
6. How does the work breakdown structure contribute to the list of activities?
7. Why must activities be mapped in a logical sequence?
8. For activity sequencing, what is the difference between "mandatory" dependencies and "discretionary" dependencies?
9. What are "external" dependencies in activity sequencing?
10. What information do network diagrams convey to the project team?
11. What are some of the possible relationships for activities in a network diagram?
12. Why is the "finish-to-start" relationship the most common relationship for many activities?
13. What is the difference between lead time and lag time?
14. Why would the project manager attempt to shorten project duration through lead time even though this may increase risks?
15. Why is it not a good idea to keep adding more workers in a project in an attempt to reduce project duration?
16. Is it correct to say that the project duration is estimated from the project schedule?
17. How are activity durations estimated?
18. How would the use of different resources with different capacities affect activity duration?
19. What is the meaning of using expert judgement to predict activity duration?
20. Why is it necessary to factor in reserve time for estimating activity duration?
21. What are some of the assumptions made to estimate activity duration?
22. Why is the project schedule iterative in nature?
23. What do you understand by the "Start No Earlier Than" time constraint?
24. What do you understand by the "Start No Later Than" time constraint?
25. What do you understand by the "Finish No Later Than" time constraint?
26. What do you understand by the "Finish No Early Than" time constraint?
27. Why is it important to identify the critical path in project time management?
28. What is float time in the critical path method?
29. What is duration compression in project time management?
30. Is there a difference between crashing and fast-tracking project activities?
31. Why is it beneficial to adopt the resource levelling and smoothing approach in projects?
32. Why is it necessary to include schedule control as part of the project integrated change control system?
33. What is the meaning of hard logic to activity sequencing?
34. What is the meaning of soft logic to activity sequencing?
35. Can activities on non-critical paths be delayed?

Chapter 7
Project Cost Management

7.1 Introduction

a. Estimating costs of the project

Having identified the resources needed to complete the project, it is now necessary to estimate how much each of these resources cost. The sum of all costs related to all resources is the total project costs. Cost estimating is the process of calculating and computing the costs of all the resources that have been identified to be necessary to complete the project. Such resources can be identified from the project scope statement, work breakdown structure, components and activities. Cost estimates for resources can in turn be affected by quality choice, risk influence and procurement matters. Resources typically include materials, manpower and machine. In addition, costs of these resources can be influenced by management competence, methods of construction and money to fund the project. Collectively, this can be abbreviated as the 6 M's. Estimating the costs of resources can be affected by economic conditions, currency fluctuations, market competition and other variations such as those relating to availability as well as demand and supply situation.

These may collectively affect the total costs of the estimates. Upon completion of cost estimating, the cost estimates for resources are derived. These show the costs needed to procure the resources necessary to complete the project. However, costs to the builder are not the same as costs to the client. The builder adds a profit margin to his costs that will then show up as a price to the client. Apart from identifying the resources needed to complete the project, the quantities of such resources must also be determined correspondingly. The amount of resources or the duration needed to use the resources is in turn estimated using labour constants and equipment outputs. For example, labour constants show the quantities of bricks that can be laid in one hour by a team of three bricklayers; i.e. square meter per hour. The duration needed by the team of three bricklayers can then be determined to estimate the labour costs associated with the bricklaying activity. Equipment outputs show for example the

© Springer Nature Singapore Pte Ltd. 2018
L.S. Pheng, *Project Management for the Built Environment*,
Management in the Built Environment,
https://doi.org/10.1007/978-981-10-6992-5_7

volume of earth that an excavator can dig in one hour; i.e. cubic meter per hour. The duration needed to excavate the entire site can then be determined to estimate the machinery costs associated with the excavation activity.

Apart from estimating the costs for using a team of bricklayers or an excavator, cost estimating also needs to know how much each of the resources costs. For example, it is necessary to know how much a bricklayer is paid; i.e. $ per hour. Likewise, it is necessary to know how much it costs to rent an excavator for a day; i.e. $ per day or $ per use. Where materials are concerned, their rates should also be made known for cost estimating. For example, it is necessary to know how much concrete costs; i.e. $ per cubic meter. Likewise, it is necessary to know how much steel reinforcement bars cost; i.e. $ per ton. If the rates for such manpower, machine and materials are not known, then these have to be estimated from various sources such as the relevant vendors.

7.2   Cost Components

a. Four types of costs

Cost estimating takes into account the various types of costs incurred for an activity that is needed to complete the project. There are generally four different types of costs to be considered for cost estimating. The first type is direct costs which refer to costs that are directly attributed to the project and cannot be shared among other projects that are also progressing at the same time. Examples of direct costs include hotel charges, airfares, long distance telephone charges, drafting expenses to produce drawings, architectural modelling costs, etc. The second type is indirect costs which are also often referred to as overheads. Indirect costs are typically costs that can be shared among projects that are running at the same time. These include for example rentals and utilities bills for head office premises as well as software licensing fees for the entire organization. A company vehicle that is shared among different projects is another example of indirect costs to the organization. The third type is variable costs which as the name suggests, can vary depending on market conditions. For example, costs of materials can vary depending on the demand and supply situation in the marketplace. Such a situation can occur during an economic recession where supply may exceed demand, causing prices of materials to come down.

On the other hand, supplies from overseas may be severely curtailed because of trade restrictions which then cause prices to increase. The fourth type is fixed costs which as the name suggests, will remain the same throughout the project. For example, the professional lump sum fees paid to the design consultants are fixed even if there is a delay in project completion. Similarly, the annual retainer fees paid to a law firm are fixed for that period of time regardless of the number of consultations made by the project organization.

The cost estimates built up from the four types of costs can however be reduced by exploring and finding new and better ways of doing things. This is where value engineering is applied which is a systematic way in which less expensive ways are identified and used to complete the same work. For example, the project may have originally identified System A for constructing the structural frame of a building at a certain cost. The cost for using System A may be viewed by the client as being expensive and he makes a request to explore other comparable systems to achieve cost reduction. Through value engineering, the project team can study comparable System B and System C to determine if the same work can be completed within the same time period but at lower costs. Upon completion of the value engineering exercise, a more informed recommendation is then made to the client for approval.

b. Historical cost information

Cost estimating cannot do without access to and availability of historical cost information. Such information can be obtained through subscriptions from commercially available cost estimating databases which can either be available on-line or printed hard copies. Preferably, such historical information should already be available in-house. Reference may be made to recently completed project files to extract cost estimating information. The cost estimating information extracted should however be adjusted for different project conditions such as those relating to difficult site locations and prevailing market situation.

Project team members with the relevant past experience or expertise may also be consulted for the cost estimating exercise. Such inputs from team members would be useful for making adjustments to cost estimates extracted from past completed project files. This is especially so when project team members have the institutional memory to provide the lessons learned for the current project that can affect cost estimating. For example, team members might have faced difficulties operating a particular system which are known only to those who have had experience working with it before. Such difficulties can continue to have implications for cost estimating and should therefore be avoided in the current project.

7.3   Approaches to Estimating

a. Analogous estimating

The analogous approach to cost estimating is based on historical information of past completed projects that are quite similar to the current project. In analogous estimating, the actual costs expended in past completed projects are taken as the platform to build on for the current project. However, not all projects are exactly the same or similar. In adopting analogous estimating, it is therefore important to consider the scope, complexity and magnitude of the current project when comparing it against the past completed projects. Other variables or specific attributes of the current project should also be flagged in analogous estimating to account for differences. For example, the time available for completion in the current project may be shorter than past completed projects because of specific client's requirements. Shorter completion time available can be translated to mean more and costlier resources are needed for a shorter period of time. It can also mean that productivity may be affected because of site congestion, thus impacting on costs adversely. Identifying such variables between past completed projects and the current project may need to depend on the expert judgement of team members.

Analogous estimating is also typically a top-down approach where senior management takes on a more active role in the cost estimating exercise. Such a top-down approach is useful for deriving quick estimates or ball-park figures requested by the client. But such an approach is also correspondingly the least accurate for its lack of attention to details. Analogous estimating provides a unit rate based on the price of a material. For example, at the micro level, the composite unit rate for floor finish using non-European granite tiles of medium quality including screed backing is approximately $200/m^2. At the macro level, the construction costs for flatted factories can range between $1100 and $1500 per square meter. The construction costs for a terrace house are estimated to be $800,000 per unit depending on its built-in area.

b. Bottom-up estimating

This is a detailed and comprehensive approach to cost estimating where each and every components of a project are accounted for in building up the project sum. Bottom-up estimating starts with the work breakdown structure and works through all its components and activities to produce the cost estimates. This approach fundamentally starts from zero base and can be a time-consuming process in having to account for everything needed to complete the project. It is however also the most accurate approach to cost estimating. The decision to use analogous estimating or bottom-up estimating depends on the stage of the project and the intended use of the estimates by the client. Analogous estimating is useful at the early stage of a project to provide an indication to the client if the project sum is likely to fall within his budget and if the project should proceed. Bottom-up estimating is useful to provide a more definitive indication of how much the project is likely to cost. In addition, details provided through bottom-up estimating can serve to check if the bids submitted by vendors are reasonable as well as for facilitating progress payments in the future.

c. How costs of resources are determined

Before cost estimating can commence, it is necessary to understand what the project includes. This is shown using the work breakdown structure, components and activities. Once details of the necessary activities are known, the resources needed to complete these activities can be identified. Hence, possessing the domain knowledge of the industry in which the project manager is in is crucial. If a project manager works in the construction industry, he needs to know about building technology and understands the processes in which the structure is built. Essentially, this has to do with thinking and is a planning process to identify the resources needed and to determine their respective costs. The planning process first identifies the 6 M's needed for the project; namely management, methods, money, materials, manpower and machinery. It then progresses to determine the quantities of these resources required, when these are needed and should these be purchased from vendors or be made in-house. Decisions can be made to purchase the resources from vendors if this is deemed to be more cost effective.

This can also happen if such resources are not available within the project organization. While ascertaining the bids or quotations from vendors, it is also necessary for the project team to put in place a process to evaluate their price and quality. The amount of resources needed may also be affected by effect of the learning curve. For projects with repetitive activities, workers can become more productive as they learn to do the same thing over and over again. As a result, costs may decrease as more of the same units are being built. The positive effect from the learning curve should also be factored in when evaluating the cost estimates for these repetitive activities. It is also a good practice to set aside contingency funds to cater for unforeseen situations such as dealing with new risks that can lead to cost overruns. Depending on the nature of the work, the reserve funds set aside can typically range between 5 and 10% of the total project sum.

d. Analysis of cost estimating results

When the cost estimating exercise is completed, checks should be made of the actual cost estimates computed for the resources needed for project completion. These checks should include an analysis of the costs apportioned to each of these resources. The analysis should be detailed enough to include all resource types whose costs can be categorized into manpower, materials, machine, computer hardware, programming software, overseas travel, provisions for inflation, etc. For example, resource types and their corresponding costs for completing a brick wall would include the bricklayers, bricks, cement, sand, water, hoisting, transportation, wastes and other overheads. In addition, where risks have been identified for an activity, the costs of the risks along with the costs of the corresponding risk responses should also be included in the cost estimate results. Hence, costs associated with the risk responses such as insurance premiums and warranties should be included. Analysis of the cost estimating results therefore provides a check whereby cost omissions if any can be identified for corrective actions to be taken.

e. Updating cost estimates

More details should be available as the project progresses. The availability of such details means that the cost estimates computed earlier should be refined and updated. Such details can include more definitive specifications relating to quality of floor tiles or the confirmed availability of a mobile crane from the pool of plants maintained by the organization. Cost estimates can be phased in at three different stages of a project wherein the magnitude of variance is different at each stage. Firstly, when the project is first initiated, little or no details are available. This typically happens at the initiating stage of a project where the project scope is still hazy. Hence, there is strong dependence on a top-down approach to provide ballpark estimates of rough order of magnitude where the variance can range from anywhere between −30 and +70%. At the early initiation stage of a project, the variance for the cost estimate is generally wider for the project team to be able to err on the safe side. This variance means that the estimate can be 30% less than or 70% more than the cost estimate. The variance is higher at 70% on the positive side so that the client is prepared to pay up to 70% more than what was estimated.

Secondly, when the project progresses to the early planning stage more details are made available. The ballpark estimate is now refined and the budget estimate is now prepared from top-down with a variance that can range from anywhere between −15 and +30%. The margin of error is now narrower and there is less room for the project team to err. This variance means that at the planning stage, the estimate can be 15% less or 30% more than the cost estimate. With more details provided at the planning stage, the client is now better prepared to accept that he needs to pay 30% more than what was estimated earlier. This is a good development compared to the 70% indicated at the initiation stage. Thirdly, as the project progresses further into the late planning stage, even more details are now available. The estimates produced at this stage are the most accurate because the availability of details allows bottom-up estimating to be carried out. In this case, cost estimates are built up from first principles taking into account the direct costs and indirect costs associated with all resources used for the components and activities.

At this point in time, the project scope is better defined and the quantities of work involved should already be known. As a result, the variance is lesser and can range from anywhere between −10 and +15%. This means that the client knows that he only needs to pay 15% more than what was estimated earlier. Likewise, the client should also know that the cost savings if any should be about 10%. As a rule of thumb, less cost variance is to be expected as more details are made available as the project progresses.

f. Documenting details of cost estimates

The cost estimates are built up based on the information and details gathered. For clarity and for records purposes, such information and details should also be documented for future reference. The documentation is useful to show how the estimates were derived. Information on the scope of work should be documented as this forms the basis for starting the cost estimating exercise. The document should also include information relating to the methods adopted for the project and the resources used. As the project progresses, cost estimating moves from a top-down approach to a bottom-up approach. While the approach to cost estimating moves from top-down to bottom-up, information on the approaches used to develop the cost estimates should be documented. All vendor quotations should form part of the information to be documented.

Where details are not yet available when cost estimating starts, the assumptions made as well as the constraints faced with respect to the methods used to complete the project must be highlighted as part of the supporting details in the documentation. For example, if it is opportune to use the services of a subsidiary company because of cheaper prices, such a decision should be documented for future reference. Assumptions such as forecasts of inflation rates and constraints such as the non-availability of supplies in the near future should be annotated systematically as part of the documentation exercise. When cost estimating faces a variance, information on the range of variance expected should be provided. For example, when it is unclear how much should be set aside for door installation in a project, a cost estimate may initially yield an amount of $200,000 plus and minus a variance of $30,000.

7.4   Cost Management Plan

a. Developing the plan

The cost management plan is developed following completion of the cost estimating exercise for the project. This is an important plan for the project team to understand how much will be spent for which portions of the project and when. This is an important aspect because cash flow is the lifeblood of the industry. The project team should ensure that positive cash flow is maintained throughout the entire project duration. The cost management plan provides detailed costs information for variances from the project costs to be managed. The plan facilitates monitoring and control as part of the project management processes.

Supporting information in the cost management plan should also include existing organizational policies and procedures on how variances are to be treated. For example, there may be an existing policy that stipulates top management approval must be obtained if the cost of a variance is above $25,000. Hence, if the cost of the variance is $30,000, the project manager must prepare a variance report with the supporting reasons for submission to top management for consideration. A meeting may then be held with top management to discuss the change request and explore alternatives. The meeting outcome may lead to approval or the initiation of an audit to study the alternatives further. The cost management plan also serves to facilitate cost budgeting. The cost budgeting process assigns a cost figure to an individual work package or costs to the project works with a view to facilitate measurements for performance. This is often heard when the project team asks "what is the budget and is the work completed within the budget?"

In essence, this is part of monitoring and control in project cost management. Fundamentally, cost estimates show the project costs by category such as the labour, machine and materials costs for constructing the foundations. On the other hand, the cost budget shows the costs across time and is used for cash flow projections. This then determines if there is a need to inject additional cash for project financing. Across the project schedule, it is therefore necessary to find out what are the cost baseline, the presence of variances if any, the costs to date and the budget at completion. Such information allows the project team to monitor spending to prevent a negative cash flow situation.

7.5   Project Budget

a. Developing the budget

The total project cost is aggregated from the costs of each work package. In developing the project budget, it is necessary to set aside a contingency amount as a reserve to provide for unforeseen events. At the broader level, the project budget may first be determined using parametric estimating to tentatively extrapolate what the costs of a project are likely to be. Parametric estimating can take the form of $ per hotel room or $ per apartment. The total project cost computed using parametric estimating allows the client to assess if he has the means to fund the project until completion. If he lacks the funds, a negative cash flow situation will occur if he is unable to reconcile this with his funding limits. This is because the project cannot have all the funds at once and the projected cash flow needs to be managed against the project deliverables based on a predetermined schedule.

Cash flow is also a concern of the contractor who needs to juggle between income and expenditure before he receives progress payments on a monthly basis. A more accurate approach to planning cash flow is through bottom-up budgeting which starts from ground zero and requires that all work packages be accounted for during the entire project duration. Hence, there is a relationship between the budget and time. At the initial stage of a building project, budgeting is for the foundation works, before moving on to superstructure works, architectural works and finally mechanical and electrical works. Bottom-up budgeting enables the budget for each stage of the building project to be determined. As projects become larger, budgeting is also expected to be more complex. For such projects, commercially available software can be used to manage the budget over time.

As the budget is monitored and controlled over time, it is also necessary to establish the cost base-line to ensure that a healthy cash flow situation can be maintained. The cost base-line in most building projects typically follows an S-curve. The S-curve allows the project team to forecast when monies need to be spent, how much need to be spent and over what period of time. Likewise, for large projects involving several phases of work, it may be necessary to establish several cost base-lines or S-curves for planning purposes. The first cost base-line is an S-curve for the entire project. If the project has three different phases, then it would be necessary to create three S-curves; one for each phase of work. Collectively, the cumulative total of all three S-curves at the phase level would sum to the S-curve for the entire project. Essentially, what this means is that the project costs for Phase 1, Phase 2 and Phase 3 would add up to the overall project cost.

7.6 Cost Control

a. Implementation issues

Project cost management needs to factor in changes arising from variations or unforeseen circumstances. Cost control is an important function of overall project cost management to ensure that the project team maintains a tight rein on costs. To do so, cost control focuses on enabling costs to change without any adverse effects on the project. As part of the project management processes of monitoring and controlling, cost control is about allowing or preventing cost changes to take place. Changes are generally unavoidable in projects. When a change does indeed occur, it must be documented in a variance report. The report sets out the background and reasons for the change as well as how this is likely to lead to a cost variance. The report should also document the advantages as well as disadvantages associated with the change. Cost control relies on several inputs for implementation.

Cost control starts with understanding the cost base-line or S-curve for the entire project duration. It then tracks the project funding requirements which in the case of building works include the monthly progress payments received by the contractor or the staged payments received by the property developer from the purchasers of a new condominium project. Cost control also tracks regular performance reports. For example, if the quality performance report is acceptable, payment is then made to the contractor. Change requests and the corresponding effects on costs and budget should also be tracked through cost control. Finally, the cost management plan is an important platform for cost control as the plan shows the overall picture of how and when costs are distributed in the project. For example, the cost management plan will show the amount allocated for the installation and commissioning of the escalators in a new shopping mall project. The plan will also indicate the start and end dates for the escalators.

b. Focus of cost control

In essence, cost control serves to ensure that there is no cost overrun and that the project spending is kept within the budget over time. Cost control therefore focuses on several issues. Firstly, if a change request occurs, cost control reviews the causes for the change and determines if the change is really necessary. It adopts a preventive approach to ensure that only genuine change requests are approved. Cost control monitors adherence to the cost base-line or S-curve. If changes are made, these will be controlled and documented as and when these happen.

Cost control therefore maintains a record of the changes made. Fundamentally, cost control is about monitoring and controlling changes in the project to minimize any adverse influence on costs. The process of cost control is also an attempt to understand and recognize cost variances. It performs cost monitoring and controlling for the client to ensure that he is not paying too much or too little for the project over its entire duration. In the event that changes are inevitable, the corresponding influence on the project must be communicated to all stakeholders. For example, there may be changes to the construction methods because of new regulatory requirements. In such a situation, cost control will strive to bring the costs affected to within the acceptable range of the stakeholders.

c. Cost control change system

It is useful for the project team to set up a cost control change system to track and document cost changes. The system should first be described using a flow chart so that all stakeholders are clear on what to do in the event that there is a change request leading to cost change. The flowchart should start with the initiation of a change request and information of how this request is likely to affect costs. These two documents then enter a tracking system which spontaneously does two things. Firstly, all documents relating to the change request, its causes, cost implications, stakeholders affected and the decision should be filed for records. At the same time, the tracking system determines if the change has already occurred.

It should be noted that in the event of an emergency, there is little or no time to seek approval for a change request when public property, health and safety is at risk. For example, an emergency of this nature can arise when flash floods occur and the project team needs to take immediate measures to minimize damage to on-going project work. When such a scenario happens, the cost change has already happened for good reasons. The documents supporting the cost change are then documented for records. On the other hand, if the cost change has not already happened, then the change request is routed for approval. At this stage, whether the cost changes are approved or not, the documents associated with either decision is then routed to the tracking system for record purposes.

7.7   Measuring Performance

a. Earned value management

Project cost management needs to measure performance to determine if the work completed is acceptable before payment is made to the vendor. The evaluation of project performance is facilitated through the concept of earned value management which measures performance of the project work against a pre-determined plan in order to identify variances. The concept is therefore useful to predict if variances are likely to occur and to determine what the final costs are likely to be when the project is completed. Earned value management starts with comparing the work completed or performed against the work planned earlier. Through the comparison, it also determines the actual cost of the work completed or performed. Project performance measurement using earned value management is an important component of cost control.

This allows forecasts to be made of future variances when compared against the expenses paid to-date. In most projects, the three common indicators of performance are Planned Value, Earned Value and Actual Cost. Firstly, Planned Value reflects the planned value of the work to-date. Secondly, Earned Value indicates the value or worth of the project to-date. Thirdly, Actual Cost refers to the cost of the project work to-date. For example, a project has a Budget at Completion of $100,000. The project is spread over 12 months. Mid-way through the project at the 6th month, the Planned Value should be $50,000. The Planned Value reflects what the project should be worth at this point in time in the project schedule. If the project is 25% complete, then the Earned Value is $25,000. However, if at the 25% mark, the Actual Cost is $27,000, then there is a negative variance of $2000. This means that the contractor has actually spent $27,000 for 25% of the work whose Earned Value is expected to be $25,000. This is not a desirable outcome.

7.8 Uncovering the Variance

a. Why this is important?

Effective cost control requires a variance to be uncovered even before it happens. It is important to uncover the variance so that appropriate preventive actions can be taken. A variance or a change is any outcome that is not the same as what was originally planned for or expected by the stakeholders. For example, a project delay is a variance because there are cost implications. For cost control purposes, a variance at the end of a project is the difference between the Budget at Completion and the Actual Costs. If Budget at Completion is $100,000 and the Actual Costs spent are $120,000 then the project variance is at a negative value of $20,000.

On the other hand, the Cost Variance can also occur during a project. The Cost Variance is the difference between the Earned Value and the Actual Cost. For example, a project has a budget of $200,000 and has completed 10% of its project value. Hence, Earned Value is $20,000. However, because of unforeseen events, $25,000 was actually spent to complete that $20,000 worth of work. Actual Cost is therefore $25,000. In this case, the Cost Variance is at a negative value of $5000 which is not desirable. With this information in mind, the project team should therefore take appropriate measures to ensure that the Cost Variance moves out of the negative range as the project progresses.

b. Schedule Variance

It should be recognized that project cost management is also closely intertwined with time. This is because if there is a project delay, costs and revenues are also likely to be affected. It is also equally true if the project is ahead of schedule. The difference between where the project actually is at the present moment and where the project was earlier planned to be is referred to as the Schedule Variance. Take for example a building project that is expected to be completed in two years with a budget of $200,000. If the project manager plans to have 60% of the work completed in the first year, then the Planned Value is $120,000.

However, for various reasons, the amount of work completed at the end of the first year is only 50%. Hence, the Earned Value for the project at the end of the first year is only $100,000. The Schedule Variance measures the difference between the Earned Value and the Planned Value. In this case, the Schedule Variance is—$20,000 (i.e. $100,000 less $120,000). This negative outcome is undesirable. Based on this result, the project manager needs to identify the causes for the delay and implement corrective measures to speed up the project progress. Identifying the causes of delay and implementing the corrective measures must take place immediately after a negative Schedule Variance occurs.

c. Cost Performance Index

In managing the cash flow, due diligence should be exercised to ensure that revenue is larger than the cost. The Cost Performance Index does this by showing the amount of work a project is completing for every dollar spent on the project. For example, a Cost Performance Index of 0.80 means that it is costing the organization $1 for every 80 cents' worth of work. In this context, the project is losing 20 cents for every dollar spent. The Cost Performance Index is computed by dividing the Earned Value by the Actual Cost. Take for example a project with a Budget At Completion of $100,000. When the project is 25% completed, the Earned Value should be $25,000. However, at the 25% completion milestone, the Actual Costs spent was $31,250. This means that $31,250 was actually spent to achieve $25,000 value at the 25% completion milestone. Hence, the Cost Performance Index is 0.80 (i.e. $25,000 divided by $31,250).

This is not a desirable scenario. Generally speaking, the closer the Cost Performance Index is to 1, the better is the project performance. Nevertheless, it may also be undesirable if the Cost Performance Index is greater than 1. Checks should therefore be made to understand why the Index is greater than 1. This may happen because of front-loading or inflated estimates submitted by the successful bidder. Front-loading means that the bidder submits rates that are higher for earlier tasks and rates that are lower for later tasks of a project. However, when aggregated across all tasks, the bidder may still be able to strategically win the job with a low tender.

When works start, the successful bidder will be paid more than what he actually needs to spend. Front-loading is a concern to the project client because he is paying more than what the builder is delivering in the project. There are corresponding risks if the client pays more at the early stages of a project and soon thereafter, the builder goes into liquidation. In this case, the client suffers a loss. The Cost Performance Index may also be high if works have already been completed but the claims for expenditure have not yet been submitted or processed. This can lead to negative cash flow for the builder.

d. Schedule Performance Index

While the Cost Performance Index shows how much work the project is completing for every dollar spent, similarly, the Schedule Performance Index monitors how close the project is on or behind schedule. Like the Cost Performance Index, the closer the Schedule Performance Index is to 1, the better the project is progressing to meet planned time-lines. The Schedule Performance Index is computed by dividing the Earned Value by the Planned Value. For example, if the Earned Value is $50,000 and the Planned Value is $80,000, then the Schedule Performance Index is 0.625. This is way off the desired index of 1. If this is not managed well, the builder is likely to face cash flow problems over time. Faced with a Schedule Performance Index of 0.625, the builder should therefore examine the causes leading to this situation and take measures to enhance Earned Value in the project.

e. Estimate to Complete

Most organizations do not have all the cash needed to finance a project throughout the project duration. This is particularly true if the project duration is long. While this may be true, organizations will however need to know how much their regular financing needs are over the project duration. At any point in time in the project, the organizations need to know how much cash they have to set aside to see the project to completion. This is where the Estimate to Complete information is required for this purpose. The Estimate to Complete essentially shows how much more money is needed by the firm to complete the project.

To do so, it is therefore necessary to know the Estimate at Completion as well as the Actual Costs. The Estimate to Completion is computed by deducting Actual Costs from the Estimate at Completion. For example, if the Estimate at Completion is $500,000 and the Actual Costs is $350,000, then the Estimate to Complete is $150,000 (i.e. $500,000 − $350,000). Essentially, what this means is that $150,000 is needed to bring the project to completion at that point in time. Information relating to the Estimate to Complete is useful for the client to plan for his cash outlay over the remaining duration leading to project completion.

f. Estimate at Completion

The client needs to know how much he will be spending in the project as the project progresses. The Estimate at Completion is a projection of what the final project costs are likely to be for the client based on the experiences in the project so far. It is also an estimate of what the total cost of the project will be to the client. Experiences in the project can be measured in some ways such as through the Cost Performance Index. Together with information on the Budget at Completion, the project manager is able to compute the Estimate at Completion at a specific point in time in the project. For example, a project has a Cost Performance Index of 0.80, which is not desirable as it is not close to 1. It has a Budget at Completion of $300,000.

Hence, the Estimate at Completion can be computed by dividing the Budget at Completion by the Cost Performance Index. In this case, we divide $300,000 by 0.80 to obtain the Estimate at Completion of $375,000. It is clear that the Estimate at Completion (i.e. $375,000) is greater than the Budget at Completion (i.e. $300,000). The conclusion is that the Estimate at Completion overruns the Budget at Completion. Corrective measures must therefore be taken to reduce the cost overrun after examining where the project went wrong. Efforts should be taken to move the Cost Performance Index closer to 1. As the Cost Performance

Index moves to 1, it is safe to say that the Budget at Completion is equal to the Estimate at Completion.

g. Dealing with anomalies

As the project progresses, some aspects of work do not meet the stipulated requirements. These are known as anomalies and should be dealt with immediately to rectify the mistakes. Anomalies can happen for many reasons, not least of which is mistakes made by people. For example, a landscaping project may involve the planting of different types of trees and shrubs for a public garden. The trees and shrubs have to be planted in a specific pattern. The entire landscaping project costs $50,000. However, about mid-way through the project, it was discovered that the trees and shrubs were planted in the wrong order. The mistake can be rectified by replanting.

However, this would require additional time and manpower to do so. At this point in time, the Earned Value of the project is $20,000. The Estimate to Complete can therefore be computed by deducting the Earned Value (i.e. $20,000) from the Budget at Completion (i.e. $50,000). The Estimate to Complete is therefore $30,000 which is how much more money is needed to see the project to complete. However, the Estimate to Complete does not include the additional money needed to rectify the mistake. Rectifying the mistake or anomaly therefore adds costs to the landscaping project. The additional time and manpower needed to correct the mistake adds to the project costs over and above that of the Estimate to Complete. It should also be recognized that dealing with anomalies do not only involve costs. At times, anomalies also affect the reputation of the company doing the work which is worse than the additional costs incurred.

Other anomalies may involve overestimating worker abilities and competence. In this context, the project is not only delayed but the workmanship may also be shoddy. Finally, at the end of the project, it is necessary to compute the Variance at Completion to assess the project cost performance. The Variance at Completion is computed by deducting the Estimate at Completion from the Budget at Completion. For example, if the Budget at Completion is $300,000 and the Estimate at Completion is $325,000, then the Variance at Completion is—$25,000 which is in negative territory. This is not desirable as the negative variance affects the client's total project budget and cash flow towards the end of the project.

7.9   Corrective Actions and Updating

a. Corrective actions

As the project advances, monitoring the health of the project budget is important. This takes place through Earned Value Management which uses a set of rules. First and foremost, Earned Value Management should always start with and focus on the Earned Value. Healthy Earned Value contributes to positive cash flow. From the various accounting conventions presented above, it should be noted that variance means deduction and that negative variance is bad for the project. Likewise, index means division and an index that is less than 1 is also bad for the project. Information is now available from the computations of variance and index.

These provide opportunities for the project manager to monitor, control and update the cost estimates as more details become available. Part of the cost monitoring and controlling process may require changes in the resources used, adopting different ways to do the work as well as changes in the sequence of events for the project. Consequently, these may also lead to changes in the cash flow projections. Applying appropriate corrective actions in a timely manner is critical for project cost management.

b. Budget update

The closing process in project cost management requires updating the budget for closure when the project is completed. The budget at closing may be different from the original budget when the project first started. The budget at closing is more often than not larger than the original budget for several reasons. Firstly, when the project is first initiated, it is often the case that not many details are available for cost planning to take place. As more details are added, this may lead to an increase in the project scope. This correspondingly leads to cost increase as well as a longer time needed to complete the project.

For this reason, it is also prudent on the part of the client to set aside a reasonable amount of contingency sum to cater for unforeseen events that may inflate project costs. This is especially true for large construction projects spanning over many years before nearing completion. When the project is finally completed, the project manager should identify useful learning points that the organization can file for future reference. These learning points can relate to the cost estimates of unique components, best practices as well as mistakes that should not be repeated in future similar projects. In the final analysis, in project cost management, expenditure must be planned for and closely monitored for the project to complete on budget.

7.10 Revision Questions

1. What information is needed for cost estimating?
2. What is the difference between cost estimating and pricing?
3. What are direct costs?
4. What are indirect costs?
5. What are variable costs?
6. What are fixed costs?
7. How does value engineering contribute to cost estimating?
8. Are organizational process assets (i.e. historical information) important for cost estimating?
9. What is analogous (from the word analogy) estimating?

10. What is bottom-up estimating?
11. How are costs of resources determined?
12. What is the possible effect of the learning curve on cost estimating?
13. When analyzing vendor bids, what is the role of the Price-Quality Method?
14. Why do projects need to factor in contingency sums?
15. What are the consequences if potential risks are not considered in cost estimating?
16. Why should estimating accuracy improve as the project progresses?
17. Why is the method statement (with supporting project details) crucial for cost estimating?
18. What is the purpose of the cost management plan?
19. What does it mean when someone says "cost budget shows costs across time"?
20. Why is it important for the project manager to pay attention to "funding limit reconciliation"?
21. Why is the cost baseline usually shown using the S-curve?
22. What is a cost variance report?
23. What does cost control focus on?
24. Why is Earned Value Management essential for project cost management?
25. What is the relationship between Planned Value, Earned Value and Actual Cost?
26. How would the reordering of event sequence and resource changes affect the cost estimates?
27. What are some of the corrective actions that may be taken for cost control?
28. Why is there a relationship between project scope and project cost?

Chapter 8
Project Quality Management

8.1 Introduction

a. Meaning of quality

Quality is a fascinating concept because it can mean different things to different people. The quality concept is rendered even more complex as it applies to tangible products and intangible services. Quality must always be looked at in sum total. This is because even a simple flaw in a product or service will singularly affect the user enjoyment of that product or service. For example, a café that serves very good food can be marred by poor or rude services. For this reason, quality is often defined as all the characteristics of a product or service that bears on its ability to satisfy implied or stated needs. Stated needs are needs explicitly mentioned by the customer. Implied needs, although not expressly stated, are implicitly expected by the customer. An example of an implied need is that of a hotel guest who expects to have a good night sleep without disturbance.

Because quality can be a difficult concept, the question is often raised as to what constitutes a level of quality that is acceptable? For example, Mercedes Benz adopts the slogan of "the best or nothing." Cars produced by Mercedes Benz therefore come with a premium where price is concerned. On the other hand, it is a known fact that cars produced by the Toyota Motor Corporation are generally cheaper compared to cars produced by Mercedes Benz. Nevertheless, despite the price differential between Mercedes Benz and Toyota, the quality of the latter is not expected to be any less desirable especially where safety is concerned. Apart from cars, price differential also exists in the food industry. Two restaurants may differ in their grades and hence their prices. The customer can enjoy the quality experience if he is willing to pay for it. The quality experience lies in the expected level of service, food and ambience. Similarly, the different grades between two restaurants do not mean that the food in the lower-grade restaurant is not edible.

© Springer Nature Singapore Pte Ltd. 2018
L.S. Pheng, *Project Management for the Built Environment*,
Management in the Built Environment,
https://doi.org/10.1007/978-981-10-6992-5_8

b. Grade and quality

In managing quality, it is important to recognize the role played by grade in the quality equation. To reiterate, quality is the sum total of the characteristics of a product or service that enable it to meet the expectations and demands of the customer. Quality fulfils the customer's expectations. On the other hand, grade is about a rank, classification or category given to products or entities with the same functional use but having different technical attributes. For example, in the construction industry, there are different grades of concrete and different grades of paint. In the service industry, there are different grades of hotels such as five-star hotels and three-star hotels.

Likewise, there are different grades of travel in airliners such as Business class and Economy class. Having made the choice, low grade may not be a problem to the customer and may be completely acceptable. However, regardless of the grade or class chosen, low quality associated with the selected grade or class will always be a problem. If a customer has a budget constraint, he can choose a lower grade product which is completely acceptable to him. But in using that lower grade product, he should not be deprived of an acceptable, minimum standard of quality. Grade and quality is therefore a total experience to the customer.

8.2 Quality Management

a. The implementation approach

There are many approaches to quality management. However, by far, the most commonly adopted approach to project quality management is anchored on ISO 9001 quality standard and guidelines issued by the International Organization for Standardization based in Geneva, Switzerland. The ISO 9001 quality management system is based on an international standard that is used throughout the world. An organization that is certified to the ISO 9001 standard in Beijing adopts the same international standard as another firm that is certified in London. Adoption of the same quality standard internationally helps to break down trade barriers and allows free flow of trade globally among countries. Correspondingly, the national standards body of a country will adopt the same ISO 9001 standard for implementation. Fundamentally, the ISO 9001 standard is an international standard that facilitates organizations to identify and follow their own quality procedures.

The standard is not a quality system per se. Instead, it is a method for developing internal procedures for an organization to follow. In developing the internal procedures, organizations should first understand what needs to be done to meet customer satisfaction. Customer requirements can be described in different ways. Firstly, customer satisfaction can mean conformance to requirements or specifications. For example, a condominium building project may specify the use of timber flooring in the bedrooms. Once the timber flooring has been installed for the bedrooms, customer satisfaction is deemed to have been met. Secondly, customer satisfaction can also mean fitness for purpose. In the same condominium building project, the use of marble tiles as the flooring in bathrooms and wet areas may be aesthetically pleasing to the eye. However, marble tiles can become slippery when wet, thus posing a danger to the occupier. Hence, in this context, the use of potentially slippery marble tiles in bathrooms is not fit for purpose.

Installations that do not met conformance to requirements or fitness for purpose should be prevented. Otherwise, these need to be corrected. Where quality is concerned, prevention is always better than correction as it is more cost-effective. In addition, corrective works may tarnish the reputation of the organization which makes the mistake. It is also management responsibility to ensure that corrective work is avoided or minimized as much as possible. At the very least, management should ensure that proper working tools and training are provided to workers for them to do their work correctly first time all the time. Defective or incorrect works can be avoided through proper planning. In this context, Dr. W.E. Deming's model can be adopted here. The model goes through an iterative "Plan-Do-Check-Act" loop to ensure that everything is in good order as planned. Deming's model is iterative in nature for the purpose of seeking continuous improvement in the work process.

Continuous improvement is also part and parcel of the ISO 9000 quality management system. The system requires organizations to measure, analyze and improve their processes and operations. The rationale is that if an operation cannot be measured, it cannot be improved. Measurement allows the analytical or thinking process to kick in for improvement to take place. The potential improvement should not however be too costly so much so that customers cannot afford to buy. This is where marginal analysis is used to evaluate the costs associated with incremental improvements to be made to a product or process. The cost is then compared with the potential increase in revenues made possible by the incremental improvements.

The four key terms used in marginal analysis are the cost of quality, quality improvement, revenue and optimal quality. What this means is that an increase in cost of quality leads to quality improvements that can potentially increase revenue. However, this is generally possible up to a point where the optimal quality is reached. Once optimal quality is reached, revenue is not likely to increase substantially even with increase in cost of quality. Cost of quality refers to expenditure spent on raising or maintaining quality standards. These include expenses for staff training, safety precautions, mitigation measures to prevent poor quality through for example closer supervision, etc.

8.3 Preparation for Quality

a. Issues to watch out for

Quality can mean different things to different people. In large projects, quality issues can span across innumerable processes and operations. It is for this reason that quality needs to be planned in, and not inspected in. Planning for quality is certainly more cost-effective as this helps to prevent mistakes, the need to correct errors and re-doing the work. Inspection of the work results is considered to be an unnecessary step if good quality can be assured. However, inspection is still required if it is mandated as part of a regulatory framework to safeguard public health and safety. Planning for quality also considers the associated cost of attaining the expected level of quality to meet conformance to specifications or fitness for purpose. Poor planning for quality gives rise to cost of non-conformance such as the expenses related to reworks and repairs.

While cost of non-conformance can generally be measured, others such as tarnished reputation and endangering public health and safety may adversely affect the organization irreparably. The project manager should therefore weigh the cost of quality with the cost of non-conformance. The spectrum of cost of non-conformance is wide, ranging from reworks, delays, material wastes, dangerous work environment, poor staff motivation, tarnished reputation and potential loss of customers. The repercussions caused by non-conformance should not be taken lightly. Avoidance of non-conformance should be built into the quality policy of the organization. This should be anchored formally using a structured approach such as the ISO 9000 quality management system, Total Quality Management or the Six Sigma methodology.

With this in mind, the quality planning process starts with understanding inputs from the project scope statement. The project scope statement spells out what is included in and what is excluded from the project. This allows the project manager to focus only on things that are necessary for quality planning. The project scope statement is reviewed comprehensively to draw out the relevant details for quality planning. For example, in the construction of a condominium building project, it is necessary for the project manager to know its functional requirements, specifications as well as building details to formulate the quality plan. Documents such as drawings and specifications will need to be reviewed and evaluated to extract information for quality planning. For example, if granite floor tiles are used in the living room area of the apartment, it is necessary for the project manager to obtain information on the quality of these tiles, their typical sizes, surface texture, colours, method of laying and country of origin, etc. Such information will go into developing the quality plan for granite floor tiles used in the living room.

At the same time, it is also necessary to review the standards and regulations that are applicable to the project. For example, there may be codes of practice relating to how electrical cables, water pipes and ducts for air-conditioning are installed in buildings. Such codes of practice should likewise be reviewed and their provisions accounted for in the respective quality plans. While relevant information is being retrieved and reviewed for quality planning, the project manager should also take this opportunity to implement the Plan-Do-Check-Act approach to determine if operational improvements can be met. This process should be documented as it is an important project management outcome for future reference.

b. Cost benefit analysis

When quality planning takes place, it is also an opportune time to assess if there are better ways to doing things. This assessment is typically based on a comparison of the benefits and costs associated with the proposed change in method of working. The cost benefit analysis is therefore a tool to aid in quality planning. The outcome should be that the benefits outweigh the costs. The benefits derived from quality management activities are that no mistakes are made, no reworks are necessary, costs are kept down, reputation is enhanced and the organization becomes more competitive. There is a close relationship between quality, productivity and competitiveness. When work is done right the first time all the time, zero defects is achieved.

To reap these benefits, it is necessary to put in place appropriate quality management activities. Costs are expended for these activities. Known as cost of quality, these activities include staff training, close supervision, provision of hand tools and personal protection equipment to enhance safety. While it is essential to invest in cost of quality in order to reap the benefits of quality management activities, this should not go overboard, leading to overkill. The project manager should recognize that efforts to deliver a quality level beyond what is demanded by the client, will cost the project additional monies. The cost of quality to be expended should therefore be aligned closely with the level of quality expected or demanded by the client. If extra features are provided that are not expected or demanded by the client, he is not obliged to pay for these additional features. If this mismatch happens, it will affect the project schedule as well as drive up the project costs. Any extra features beyond the client's expectation which he is not paying for is known as gold-plating. However, not every additional feature that is part of quality management activities amount to gold-plating. This depends on the risks involved and an assessment if such risks can be accepted. For example, additional precautions must be taken if deep excavation work is to be carried out in poor soil conditions near to existing old heritage buildings.

In such a scenario, the advice of a geotechnical engineer should be sought on the safety precautions to be taken. Such advice may include the use of special excavation methods that cause minimum vibration and the installation of geotechnical sensors around the site to continuously monitor surrounding soil conditions. These require expenditure but are not classified as gold-plating. On the other hand, if the new building site for another project is prone to flooding, the builder will need to put in place measures to mitigate floods during the construction phase. The builder should be able to do this by installing a temporary drainage system around the site with sufficient gradient to drain excess water away quickly. For this purpose, at the construction phase, the builder needs not engage the services of a landscape architect to produce an elaborate drainage design which is only temporary in nature. Engaging the costly services of a landscape architect would amount to gold-plating in this case.

c. Benchmarking

In examining the results of the cost benefit analysis, it is also necessary to compare the computed costs and benefits with other similar projects to determine if the decision leads to value for money or is a best practice. This comparison is carried out through benchmarking which is also a tool for quality planning. Fundamentally, benchmarking is about comparing the practices of one project with those of others. Benchmarking requires measurements before comparison can be made. These measurements can be at different levels. At the industry level, comparison is made against industry standards. For example, the company may compare its quality achievements by comparing its quality scores with the industry average quality scores under the Construction Quality Assessment System which is the national yardstick for assessing workmanship quality in Singapore.

At the company level, comparisons may be made with the pricing levels of competitors or their performance. Key performance indicators can be identified for comparison between companies. Such indicators can include the bid prices, time required for project completion and productivity of a roofer to complete installing one square metre of clay roof tiles in one hour. Benchmarking allows the company to evaluate where its current performance is relative to those of its competitors. The company can then use the results from the evaluation to determine where it has gone wrong or done right. This allows the company to continuously improve its operations to enhance quality performance and competitiveness. In some cases, the company may need to develop proto-types for test-bedding to determine how best the project should proceed. This is pertinent in cases where many different variables can be present and there is a need to establish the best solution in terms of quality, cost and time.

Experimentation of the various design alternatives may be needed for this purpose. For example, experimentation can take place in a wind tunnel laboratory to determine the most optimal flow of natural ventilation through a new building project. The experiment can be designed with the building sited with different orientations to the sun and expected wind directions. The façade of the proposed building can be cladded with different material types, projections and openings to simulate how different designs affect cross ventilation within the building at different times of the year. From the experiments conducted in the wind tunnel laboratory, the design variables that cause unacceptable ventilation effects are isolated and adjustments made to achieve better results. While various design alternatives may be tweaked to seek the best results, it should also be kept in mind that quality should not be enhanced so much so that costs spiraled out of control.

8.4   Quality Management Plan

a. Creating the plan

The ISO 9001 standard is by far one of the most common approaches to creating a quality management system. Certification to meet the requirements of the ISO 9001 standard can take place at the project level or the company level although the latter is often the preferred choice. In adopting the ISO 9001 quality management system, the quality management plan is referred to as the project quality system. This addresses three key matters. The first matter addressed in the project quality system relates to quality control which is inspection-oriented. Through inspections, results that are unsatisfactory are identified. These unsatisfactory results are then analyzed to identify their root causes for poor performance. These root causes for poor performance are then eliminated.

Quality control in the construction industry can take place through for example the slump tests and concrete cube tests to determine the quality of concrete delivered to the site. In mass production, such as those relating to bolts and nuts, it is however not possible to inspect every single bolt and nut. In such a scenario involving large scale mass production, random inspections are made through the statistical quality control process. Quality control has also been extended to the formation of quality control circles involving small groups of employees who meet on a regular basis to brainstorm and solve quality related problems that they have identified in their work. Quality control circles have also been referred to as work improvement teams. The second matter addressed in the project quality system relates to quality assurance. This is where performance is evaluated to ensure that the relevant quality standards are met in the project.

Quality assurance precedes the current quality management system approach and was the focus of earlier ISO 9000 standards that have now been super-ceded. Quality assurance is very much a document-driven process which does not place any emphasis on continuous improvement. Being a document-driven process, quality assurance essentially requires the job-holder to say what he does, does what he says, and to record that he has done it. When the document-driven approach is taken to the extreme, quality assurance invariably creates a lot of paperwork for the organization, right down to ludicrously documenting how a computer should be switched on. In the quality assurance approach, job-holders are more interested in finding the defects and then fixing the problems. The important task of seeking continuous improvement is absent in quality assurance. This is also primarily the reason why quality assurance gave way to quality management in the ISO 9001 standard. This leads to the third matter addressed in the project quality system which is quality improvement.

However, for quality improvement to happen, it is necessary to measure and evaluate performance before corrective actions can be identified and taken. This is provided for in the ISO 9001 quality management system where quality improvement efforts are built into the system. The measurement of performance is the first step towards quality improvement. Operationally, performance must be capable of being defined and measured to facilitate data collection. Performance metrics should therefore be quantifiable with values used to measure an operation, process or work item. This may also take place within the framework of statistical process control or the Six Sigma methodology. Measurements for data collection should be concise, definitive and unambiguous. For example, a company may be adopting the Construction Quality Assessment System to quantify the workmanship quality of its projects. As the national yardstick, the industry average quality score for a specific building type such as residential construction is published by the building authorities on a yearly basis.

The industry average quality score for residential construction was established to be 78 points last year. In the current residential project, the client may wish to set a target for his building to obtain a quality score of 5 points above the industry average quality score of 78 points. In this case, the goal is to obtain at least 83 points for the new project. The project management team should therefore set in motion appropriate quality plans to ensure that this target is achieved upon project completion. Part of the quality plans can include relevant checklists to ensure that nothing is amiss when the project commences. A checklist contains a list of activities which the worker can check off to ensure that each task is completed and that tasks are completed in the right sequence.

From the quality viewpoint, the checklist serves as a useful reminder for the workers to complete all the tasks that they have been assigned. For example, a checklist for the project team to prepare for concrete casting on site may include getting the equipment ready for the slump test, cube test, concrete pump, vibrator, shovels, trowels, etc. Checklists are especially useful for safety management such as those relating to demolition work, excavation work, access to confined spaces such as manholes, electrical work, etc.

8.5   Quality Audits

a. Purposes

When a vendor or subcontractor claims that the work is completed, how would the project team know that it has indeed been completed satisfactorily to meet the project requirements? This is where quality audits come into provide that confirmation. Quality audits formally review what have been completed to discover what works and what does not work in the project. However, while quality audits may be a result of contractual requirements, these should be viewed as a learning process for improvement and not to find faults with others. The final result of the quality audit is to help improve performance for a process, a project or an entire organization.

Quality audits based on the ISO 9001 standard can be both internal and external in nature. Internal quality audits are conducted by employees in the same organization as a self-checking process. This typically takes place before the external quality audits that are conducted by independent third party auditors. The purpose of external quality audits in the context of the ISO 9001 standard is for the purpose of certification or re-certification. This is because an organization is certified to the ISO 9001 standard for only a period of time. When the ISO 9001 certificate expires, it needs to be renewed. Other than the internal and external quality audits that are conducted at regular intervals, there are also other quality audits on the work completed in the project that are conducted on a monthly basis.

These monthly quality audits are typically conducted for the purpose of certifying monthly progress payments to the builder. Apart from quality audits that are conducted at regular intervals, it is also useful for the project management team to carry out surprise quality audits without notification given to the workers. Such surprise quality audits allow the project management team to determine if short cuts have been taken by the workers. For example, a surprise check may be made to confirm that the ponding test has indeed been completed to establish the integrity of the waterproofing membrane before the laying of screed and tiles commence in wet areas such as bathrooms and kitchens.

8.6 Quality Control

a. What it does?

When production or installation works commence, it is necessary to inspect the products to ensure that these are free from errors and defects. This forms part of quality control for the project. Quality control typically focuses on the property and dimensional aspects of a product. Take for example a 2-hour fire door that is to be used in a commercial building. From testing in a fire laboratory, quality control establishes if the said fire doors to be used in the commercial building are able to withstand fire for at least two hours. Quality control also determines if the dimensions of the said fire doors are correct and that the door will fit snuggly within the door opening without any gaps that may allow fire to spread.

In mass production such as those relating to ceramic tiles, it may not be possible to inspect each and every tile that comes through the production line. In such cases, statistical quality control is implemented based on sampling and probability. Sampling and probability help to reduce inspection costs but yet is able to identify random causes to determine expected quality variances. Results of the statistical quality control exercise should however be linked to the tolerance range expected by the client. The results are compared with the upper and lower limits of the tolerance range to assess if the product meets the acceptable level of quality. The tolerance range serves as the control limits for immediate rectification actions to be taken if poor quality standards are observed from the random samples.

b. Quality control approaches

Several approaches are available for quality control. By far, the most common approach for quality control is by inspecting the results. While it may be argued that inspection is unnecessary if things are done correctly first time all the time, inspection is still needed to prove conformance with requirements for the purpose of making progress payments to the vendor or builder. There are many ways in which inspection can take place. The simplest approach to inspection is by means of a site walkthrough. This involves the project team members walking through the site to take notes of the extent as well as the quality of the work completed to-date. Inspecting results may also be facilitated through reviews of mock-ups and of the products to be used in the project.

Once these product reviews have been approved, the project proceeds to installation and inspection is deemed to have been completed successfully for the products concerned. Quality audit also forms part of the inspection process to prove conformance of the project to requirements. Following the completion of inspection, the project management team may discover slipshod quality standards in certain parts of the project. Some of the shoddy quality standards may be caused by a myriad of inter-related factors or variables. To better understand how the slipshod quality standards occur and what variables caused these to happen, a more detailed analysis can be conducted using the Fish-bone diagram. This diagram essentially shows the cause-and-effect relationships between variables involved with a process. The analysis identifies the major causes as well as the contributory causes leading to the effect or in this case, the shoddy quality standards. The analysis is similar to a mind-map that establishes the relationships between the causes that lead to the slipshod quality identified.

Quality control can also make use of flowcharts which highlight how a process flows through a system. For example, frequent change requests can affect how the project progresses and consequently quality as well if the changes are not well administered. An appropriate process or system flowchart provides a clear understanding to the project stakeholders how a change request is submitted, evaluated and decided upon through the change control system. The flowchart is quite similar to a checklist but is presented in the form of a diagram. It provides a step-by-step guide to the project management team to handle the change request. Through the flowchart, all project stakeholders know exactly what to do in order to manage and control quality. Apart from flowcharts, quality control can also make use of statistical process control charts to track the performance of a project over time. Such control charts are typically useful for repetitive operations such as those used for the large scale production of tiles, bricks and ironmongeries.

The statistical process control chart is based on the central limit theorem with a normal distribution. The peak of the normal distribution curve represents the mean. The control limits are associated with the standard deviation or sigma both ways which represent the upper control limit and the lower control limit. The quality boundaries are determined by the standard deviations from the mean, upon which the Six Sigma methodology is based. Quality standards are considered to be out of control for a product if its attributes fall outside the predetermined requirements. Once this happens, immediate actions should be taken to identify the causes for the deviation and to take appropriate measures to rectify the errors. If the control charts adopt the Six Sigma methodology, the upper control limits will be set typically at +3 or +6 sigma. Likewise, the lower control limits will be set correspondingly at −3 or −6 sigma.

The sigma or standard deviation shows the degree of correctness of a specific product or service attribute which the measurement is targeted at. For example, in the case of a product attribute, this may be related to the dimensional accuracy of a tile. Similarly, in the case of a service attribute, this may be related to the expected waiting time at a service counter. In an ideal situation, ±6-sigma only allows two defects to happen in one million opportunities which represent 99.99% correctness. At ±3-sigma, the degree of correctness is 99.73%. At ±1-sigma, the degree of correctness drops to 68.26%. It is however not always necessary to strive for 6-sigma perfection. This is because costs may increase substantially if the sigma level is enhanced.

Nevertheless, in a situation where public health and safety can be affected adversely, there cannot be any room for error. In such circumstances, 100% correctness is necessary. Hence, in considering the cost of quality, the project manager needs to balance the costs associated with obtaining perfection with the costs of say obtaining just 98% correctness. In any case, once the desirable sigma level has been decided, the upper and lower control limits are also determined. Any results that fall beyond the upper or lower control limit are then deemed to be out of control. Another useful tool for quality control is the Pareto diagram. Just like it may not be necessary to strive for 100% perfection in the Six Sigma methodology, it may also not be necessary to focus efforts on all and sundry.

Projects have limited resources. Hence the project management team may want to consider working first on the more severe or largest problems before moving on to the less severe or smallest problems. This is all a matter of prioritizing work based on the Pareto's Law which states that 80% of the problems come from 20% of the issues identified. It may also be possible that in resolving the largest problems, the smallest and less severe problems may also simply disappear. The Pareto diagram is essentially a histogram that ranks the issues and problems from the largest to the smallest. By so doing, the most urgent work can be identified for limited resources to be targeted first in the project. Preparation of the Pareto diagram may however require measurements of the extent of the issues or problems identified. Such measurements may be related to the total number of failures for a particular issue or problem.

However, given that extensive data collection costs money, statistical random sampling can help to reduce the costs of quality control. Statistical random sampling makes use of a percentage of the findings to test for quality of the population. Such measurements may also be extended further for trend analysis which provides useful information for quality control. Trending includes regression analysis, correlation and time series. It is fundamentally about taking past results to forecast future performance. However, for trend analysis to be accurate there must be sufficient past records to facilitate the forecasting of results to estimate current expectations for the purpose of quality control. Caution should also be taken to ensure that spurious relationships are not taken on board without checking for reliability and consistency.

c. Outcomes of quality control

The ultimate goal of quality control is to strive for quality improvements and performance. The outcomes of quality control lead to decisions to either accept or reject the work. Quality control should not be about finger-pointing and finding faults with others. Instead, it should be about drawing pertinent lessons to enhance quality improvements. The quality control exercise helps to eliminate reworks. This renders the project more cost effective as the work is always done right the first time all the time. In so doing, it also serves to reduce disruptions for the project to keep abreast of its time schedule.

Many tools are available for quality control. These include checklists to demonstrate that all works are completed to meet requirements without missing any items. Such checklists also form part of the project documentation for future reference. As such checks are constantly conducted when the project progresses, appropriate adjustments are made to ensure deviations do not spiral out of control. Immediate corrective actions should therefore be taken to bring these deviations back on track. However, the project manager should preferably take planned preventive actions to ensure that quality performance improves over time. The recommendations from the quality control exercise should be attended to as soon as possible so that problems do not fester with time. Such recommendations may include the need for defect repairs, corrective actions, preventive measures or change requests made to avoid the anticipated defects.

In the final analysis, the project manager should recognize that quality means delivering the work to meet the project scope and its design specifications; nothing more and nothing less! It is not about gold-plating to produce more than is necessary. It also needs to be reiterated that low grade is not always a problem so long as this is what the customer asks for. However, low quality, regardless of the grades, is always a problem with the customer. Meeting customer satisfaction means that the project manager needs to in the first instance ensure conformance to specifications and fitness for purpose are achieved in the project. Correspondingly, the project manager also needs to strive for continuous improvements in the processes and operations as part of the quality management system for the project.

8.7   Revision Questions

1. What is quality?
2. Why is it difficult to define quality?
3. Why is low quality a problem while low grade is not a problem?
4. Why is the ISO 9001 quality management system important internationally?
5. What is the role of preventive actions in quality management?
6. What is the role of corrective actions in quality management?
7. Why is continuous improvement important in quality management?
8. What is cost of quality?
9. Why is it important for projects to avoid cost of non-conformance?
10. Why is it necessary for the Plan-Do-Check-Act methodology to be iterative?

11. Is Cost Benefit Analysis necessary for quality management?
12. What are some of the situations when projects do "gold plating"?
13. How does benchmarking benefit quality management?
14. Why are metrics needed for benchmarking?
15. What are the reasons for experiments or test-bedding as part of quality management?
16. Why is quality control inspection-driven?
17. Why do critics allege that quality assurance is document-driven?
18. What is the different between quality assurance and quality management?
19. In what ways do checklists support quality management?
20. Can quality audits be avoided?
21. What does statistical quality control or SQC entails?
22. Are inspections absolutely needed in quality management?
23. In what way is the Ishikawa Diagram useful for quality management?
24. In what way is the Process Flowchart useful for quality management?
25. What are Control Charts for quality management?
26. In what way can the Six Sigma methodology be used for continuous improvement?
27. When is the Pareto's Law useful for quality management?
28. Why is statistical sampling useful for quality control?
29. When is trend analysis useful for quality management?
30. What are some of the possible outcomes after the quality control results are made known?
31. Is it strategic to deliver a quality standard that is higher than what have been specified and agreed in the project scope?

Chapter 9
Project Human Resource Management

9.1 Human Resource Management

a. What it entails?

Organizations are made up of and are run by people. In a school environment, for example, human resource management is all about people. They include the teaching staff, students, parents, alumni, members of the school advisory committee and vendors. When the school is being refurbished or upgraded, the people involved are extended to include the consultant architect, engineers, quantity surveyors, contractors and suppliers who come together to work in the project and disband upon project completion. When productivity slacks in the school project environment, the people involved may be the cause of the slowdown. Getting the right people in the project organization is crucial to get the project off to a good start. Human resource management is therefore about recruitment, selection, retention, compensations, benefits, training, development, performance management and talent management.

For an organization to thrive, human resource management is also about employee engagement and work-life balance. Human resource management is clearly multi-faceted in nature. It reflects how the project manager deals with, lead and manage the project team in the first instance. This role is then extended to include orchestrating the clients, project partners, stakeholders and other contributors to accomplish the desired outcomes for the project. Human resource management is therefore not only confined to the internal organization but is also extended to include other external organizations. In a large organization, human resource management is not only looked after by the project manager. It is also the role of the human resource department in the head office of the large organization. This is especially so where there is sharing of a central pool of human resources among different projects.

© Springer Nature Singapore Pte Ltd. 2018
L.S. Pheng, *Project Management for the Built Environment*,
Management in the Built Environment,
https://doi.org/10.1007/978-981-10-6992-5_9

9.2   Human Resource Planning

a. What it does?

Human resource planning is about matching and mapping appropriate people or groups of people with the roles, responsibilities and lines of reporting in projects. It is necessary to plan to identify the people involved and how they are to contribute to the project. Having identified the appropriate people, human resource planning determines what their roles are, whom they report to, who report to them, and how they influence the project. Human resource planning includes people both within and outside the project organization. As stakeholders, their roles within the project may also change with time. For example, there may be staff secondment from a vendor to the project organization for a period of time. Their involvement in the project should therefore be identified, reviewed and revised as the project advances.

b. Human resource at interfaces

Human resource planning should take into consideration the various interfaces within which people interact. People do not work in silos in projects. They have to work with one another across boundaries for synergy. There are five different types of project interfaces where people and groups work to complete the project. The first intersection relates to organizational interfaces which involve people working in the organization but in different functions such as those in the Finance Department, Legal Department, Human Resource Department, Marketing Department, Construction Department, Information Technology Department and Administration Department.

The employees in the different functional departments may all need to be involved with one another throughout the project or their involvement may fluctuate depending on the needs of the project over time. For example, at the start of the project, employees from the Information Technology Department are mobilized to set up the computing facilities at the project site office. After setting up, they then return to the head office. They will only return to the project site office to intermittently trouble-shoot or to dismantle the computing facilities when the project ends. The cultural differences between various functional departments should also be noted. For example, while the Marketing Department may not hesitate to spend generously to secure clients, the Finance Department is more likely than not to rein in such spending.

The second intersection is concerned with technical interfaces. This is where there is a need to synergize as well as synchronize the inputs from various technical professions. For example, in constructing a new building, the project team is made up of architects, engineers, surveyors, contractors and vendors. The outputs provided by one member of the technical team are the inputs to another member of the technical team. In this case, the architectural design provided by the architect is sent to the structural engineer who then works in the structural requirements for the building. When completed, this is in turn sent to the quantity surveyor for cost estimating to ensure that the total project sum is within the client's budget. Hence, every building professional have a role to play at the project's technical interfaces. This is aided to a large extent today by Building Information Modelling.

The third intersection relates to interpersonal interfaces which describe the relationships between people working in the project. This is also concerned with different communications modes in the reporting relationships which can be informal such as discussions that take place along the corridor or formal involving written reports. Interpersonal interfaces are also prone to conflicts for various reasons such as miscommunications or cultural differences. Such conflicts should be managed in a timely manner to prevent them from spiraling out of control. The fourth intersection relates to logistical interfaces. This is concerned with stake-holders who are involved with logistics in the project supply chain. The logistical interfaces in a project should be managed effectively for supplies to be synchronized for deliveries to the project without suppliers getting into each other's way.

Synchronization is also necessary to avoid site congestion if too many suppliers converge at the site at the same time. Logistical interfaces can be rendered complex if these involve suppliers from different countries operating with different time zones and cultural values. Finally, the fifth intersection is the political interfaces which typically take the form of office politics. Political interfaces occur when stakeholders with different goals and vested interests do not agree on common issues that may adversely affect them. Political interfaces can give rise to informal power obtained through informal alliances with like-minded stakeholders. Taken to the extreme, office politics can derail a project.

c. Staffing requirements

Projects require people to run, manage and complete the various activities identi-fied. In human resource planning, it is therefore necessary to identify the various roles needed in a project to complete the activities assigned. Take for example a refurbishment project for a school involving upgrading the air-conditioning system and the light fittings. The people needed for this refurbishment project would include the mechanical and electrical engineers, installers for the air-conditioning ducting and equipment, as well as the electricians. Depending on the size of the project, it is also necessary to determine the number of people required to complete the work to meet the project deadline. Human resource planning also considers if these people are already available within the resource pool of the project team who can be directed to work in the refurbishment project. If such a resource pool is not available, alternative sources should be explored such as through outsourcing, acquiring new employees or training current employees to take on the tasks.

d. Project constraints

Project constraints are also factored in for human resource planning. Such con-straints may be related in the first instance to the existing organizational structure which determines if the project manager has the authority to assign people to work. Different organizational structures may provide the project manager with high or low power. At one end of the spectrum, this may take the form of a project-based organizational structure where the project manager possesses high power. At the other end of the spectrum, the project manager has low power in the functional organizational structure. This is because in the functional organizational structure, the project manager reports to the functional manager.

Midway in between the project-based and functional organizational structure lies the matrix structure. In the matrix structure, employees have to report to two superiors who are the project manager and the functional manager. While resources may be shared, the matrix structure can create confusion if there are conflicting instructions sent to the employee by the project manager and the functional manager. Apart from project constraints caused by organizational structures, there are also other external constraints such as those caused by unions with their collective bargaining agreements. These agreements may create further reporting relationships which should be included in human resource planning. Other constraints can occur if there is a specific project management preference to adopt an organizational structure that proved to be successful in past projects. Preference to use the same organizational model is expressly stipulated to replicate success. For example, a repeat and influential client may specifically request that a particular project manager be assigned to the project because of past successful working relationships.

On the other hand, some past projects might have faced some difficulties because of the limitations of the project team. Faced with such a scenario, human resource planning should also factor in the likelihood that the workflow of a new project may not be as productive if the same project team runs it. In addition, for special projects such as those relating to restoration works for old buildings, additional constraints may surface because of the need for unique skills or individuals who are only available from overseas. For example, in the restoration work for an old Chinese temple, skilled artisans from Hefei, China have to be flown to Singapore for this purpose. Similarly, in the restoration work for an old church building, skilled artisans from Paris, France have to be sourced for this purpose. These are procurement constraints for which human resource planning needs to consider.

e. Completion of planning

Human resource planning considers the work breakdown structure of the project and how its various components and activities take place over time throughout the entire project duration. To avoid undesirable peaks and troughs in the project, resource levelling and smoothing is essential in human resource planning. In so doing, human resource planning factors in all constraints and opportunities identified for the project. The constraints and opportunities depend on the nature of the project, competence of the project team, client's demands and stakeholders' requirements. Nevertheless, instead of starting on a clean slate, the project manager should make reference to existing human resource planning templates for past similar projects. These templates should allow him to draw lessons from past successful projects for the current project. With appropriate modifications, the existing templates would allow the project manager to determine the various roles and responsibilities of people in the current project along with their reporting structures.

Once these are established, the human resource department should provide the following details for implementation. Firstly, the individuals should be identified together with descriptions of their roles and job responsibilities. For example, if a "roving" project manager is appointed to oversee a few small projects in order to save costs, this should be identified in human resource planning for this to take effect. The human resource department should also identify the reporting structures of individuals using appropriate organizational charts for both the head office and the project. It is also a good human resource management practice to indicate the role and autonomy of the project manager to let him know what he is authorized to decide and what he needs to seek further approval from top management. For example, the project manager may be authorized to decide on changes in the project with a value not exceeding $25,000. If a change request surfaces with a value of just $10,000, he can proceed to decide on whether to approve or reject the request.

Because projects and organizations are made up of people, policies relating to staff disciplinary matters must also be implemented as part of good human resource management practice. These policies should set out the nature of the offences and their corresponding disciplinary actions. Such actions can include reprimands, probation, warnings, and in extreme cases, termination of the service of the employee. All employees should be notified of the details of these disciplinary actions. For example, an existing site policy may relate to workers not wearing personal protection equipment. Workers may be fined $20 for the offence with the fine amount going to a common pool as staff welfare funds. It is often a practice in organizations for the human resource department to pair a junior employee with a senior employee for mentoring purposes. If there is a mentoring system in the organization, the role of the mentor must be defined and explained clearly to the employees involved so as not to cause unrealistic expectations. Likewise, if other similar terms are being used such as champion, advisor, expert, etc. such terms must also be defined and explained.

9.3 Theories of Organizations

a. Purposes of such theories

Management and organizational theories provide better understanding of how and why people behave in certain ways. With this understanding in mind, the project manager is then better able to monitor and control people in the project through more effective management of their behavior and performance. Fundamentally, a theory tells a story based on a model. Organizational theories allow the project manager to identify the strengths, weaknesses and behavior of individuals in the project. This understanding enables the project manager to provide better guidance to the project team to move the project forward. Numerous organizational theories have been promulgated over the years.

These relate primarily to how the different needs of individuals motivate them in the workplace in different ways. Organizational theories also explain why some individuals are motivated, self-driven and do not require close supervision in the workplace. This contrasts sharply with others who are not motivated and where close supervision is needed for them to perform. Organizational theories also propound different factors that affect motivation according to employee expectations and the management styles that they have been exposed to.

b. Hierarchy of needs

The concept behind the hierarchy of needs was proposed by Abraham Maslow. This posits that people work in order to fulfil a hierarchy of needs which leads ultimately to self-actualization. Insofar as human resource management is concerned, Maslow's hierarchy of needs is an important consideration for staff recruitment, motivation and retention. There are five levels in the hierarchy. Firstly, beginning at the bottom of the hierarchy are the basic physiological needs. These basic needs relate to the necessities for daily living and include air, water, food, clothing and shelter. Hence, where application is concerned, the human resource management department needs to ensure that these basic necessities are present in the workplace for employees. For example, if the project site is in an isolated location, arrangements must be made to ensure that at the very least, there is a ready supply of adequate food, water and shelter for the employees.

The second level in the hierarchy relates to safety needs where people look for security and stability in the workplace and organization culture. For example, if the project site has a poor safety record, employees will not feel secured enough to contribute productively to the tasks assigned to them. Safety measures such as the provision of personal protection equipment including safety helmets, harnesses and safety boots serve to provide assurance to the employees. Beyond physiological and safety needs, people work in organizations and interact with one another. The third level in the hierarchy relates to social needs where it is recognized that people are social creatures who desire love, approval and friendship. Hence, where human resource management is concerned, efforts should be put into promote a collegial working environment and teamwork. Sporting events, company dinners and social gatherings are some examples where employees can socialize to build rapport, trust and friendship.

With social needs out of the way, it is natural for people to seek esteem needs at the fourth level in the hierarchy of needs. At this level, people seek respect, recognition, appreciation, acceptance and peer approval. Such recognition can be accorded by an organization by giving awards to appreciate the long years of service rendered by loyal employees. Helpful employees who have gone out of their way to help colleagues can likewise be honored at an appreciation ceremony. At the pinnacle of the hierarchy at the fifth level lies self-actualization needs. This is where people seek fulfilment in what they are doing through gaining knowledge or personal growth. Organizations should therefore provide the opportunities for employees to reach their maximum potential. This can be done for example by sending them for leadership programs or tapping on their experience to serve as mentors to their younger colleagues.

c. Theory of motivation

The theory of motivation is closely associated with the hierarchy of needs in some ways. This theory was proposed by Frederick Herzberg to explain the two descriptors for success in motivating people. The first descriptor is termed expected hygiene agents where all workers are expected to have job security, salaries, safe and comfortable working environment, sense of belongingness, and collegial working relationships. The second descriptor is termed motivating agents which motivate people to excel in what they do. The attributes associated with motivating agents include responsibilities, recognition, appreciation, opportunity for further education, chance to excel, etc. Fundamentally, motivating agents are linked closely with opportunities associated with work other than just purely financial rewards. For these reasons, hygiene factors are not likely to motivate people to perform since these are already expected attributes.

Nevertheless, the absence of hygiene factors will discourage workers from performing well. In a nutshell, motivating factors must be present for people to want to excel in what they do. Motivating agents lead to high level of performance. The lack of expected hygiene agents leads to low level of performance. For workers to excel in their work, they must first be satisfied with the expected hygiene agents, exit from the influences of these hygiene agents and progress to the motivating agents which promote performance. What this means is that at some point in time, after experiencing the expected hygiene agents, financial rewards and salaries may no longer provide the motivation to perform. Instead, a simple friendly pat on the back as a form of appreciation will spur the employee to strive for excellent performance.

d. Theory X and Theory Y

The concept of Theory X and Theory Y was proposed by Douglas McGregor. The concept essentially states that there are two types of workers; the good worker and the bad worker. Theory X posits that people who fall in this category are fundamentally bad because they avoid work, avoid responsibility, are lazy, cannot be trusted and lack the capability to accomplish much in life. Hence, because of these traits, people who fall within the Theory X category must be supervised and watched closely all the time because they cannot be trusted. The lesson for human resource management is that Theory X people must be micro-managed. On the other hand, Theory Y posits that people who fall in this category are self-motivated, proactive and have the drive and ability to accomplish much in life.

For this reason, Theory Y people are able to work independently and do not require close supervision. The human resource manager should therefore try to identify Theory Y people early to nurture and groom them for the organization. Theory X people on the other hand need to be counselled and guided. Nevertheless, this is not to suggest that people will be stuck in either Theory X or Theory Y permanently. Over time, with counselling and mentoring, people may be motivated to shift closer to the Theory Y territory instead of languishing permanently in the Theory X sink-hole. For this reason, while it may be useful to distinguish two different sets of traits using Theory X and Theory Y, the temptation to stereotype people should be avoided.

e. Theory Z

The concept of Theory Z was proposed by William Ouchi at a time when the world at large was fascinated with how Japan managed to pull herself out of the doldrums of defeat in the Second World War to become one of the economic powerhouses. In short, Theory Z attempts to explain the reasons for Japan's success. This was attributed to the participatory management style of the Japanese where employees are motivated by a sense of commitment to the organization they work for. It is the belief that singular individual interests should not take precedence over what is good for the group, organization and society at large. It was also the belief that opportunities and career advancement are commensurate with commitment to the organization. At the time when Theory Z was proposed, Japanese companies typically practiced life-long employment for workers who stay with one company until they retire. This cultivated a strong sense of loyalty where employees are dedicated to their company who in turn is dedicated to them. In this context, the employees believe that the company belongs to them and the onus is on them to make sure that the company continues to strive and do well. The rationale is that if the company sinks, they too will sink along with their company.

Because workers are loyal and committed, they are empowered to make important decisions that may even result in the entire production flow in the factory to shut down in order to resolve a manufacturing problem. The problem is then resolved collectively before the production line restarts. The participatory management style also shows up in the quality control circles which help to foster teamwork among workers. While quality control circles are still widely practiced today, some aspects of Theory Z do not appear to be that relevant today. Over time, for example, life-long employment has now been undermined in some organizations because of intense global competition in the marketplace. Increasingly, projects are becoming more international in outlook with cross-border businesses involving Japanese companies. Human resource managers and project managers can still stand to benefit from understanding Theory Z propositions when interacting with Japanese companies and their employees.

f. Expectancy theory

Victor Vroom proposed the Expectancy theory to explain that the behavior of people is influenced by what they expect as a consequence of their behavior. The expected reward is what people work towards and which affect their behavior. The expectancy in this theory is the belief that if a person tries hard, he can do better. And if he does better, he will receive a better reward. What he expects and what he does is the value of the outcome to him relative to him remaining indifferent or averse to the work. In essence, outcome depends on performance which in turn depends on the effort put in. Expectancy however can be influenced by the work environment in which the individual is in. Issues such as peer support and subordinate cooperation, availability of information, tools, materials, past successes, etc. are environmental factors that influence the effort put in by an individual.

This is in turn affected by the traits, abilities and values of the individual that determine how he reacts to the situation and what he expects to get out of work. Examples of rewards expected by an employee based on the hard work he puts in include performance bonus and share options. However, expectancy of rewards in monetary terms may not always work out well in non-profit organizations such as charities, religious bodies and non-governmental organizations. In such organizations, monetary rewards are not prioritized in order to support the passion and beliefs an individual has in the work concerned. Nevertheless, to manage people well, the project manager should recognize that people who work for him expect to be rewarded for their efforts.

9.4 Styles of Management

a. Concern for people and production

Organizations are made up of people who act and behave differently. Different management styles are therefore needed to deal with different employees. Similarly, organizations have managers who practice different management styles which can be autocratic or democratic in nature. An autocratic manager is one who makes all the decisions without consultation with his subordinates. The autocratic management style appears suitable for managing Theory X employees who need to be supervised closely. On the other hand, a democratic manager involves his team members in making decisions. This management style is able to obtain buy-in from team members who have played a role in making the decisions. The democratic management style appears appropriate for managing Theory Y employees who do not need to be supervised closely.

Pushed to the extreme, the democratic management style may adopt a laissez faire approach where the employees are self-initiated and are happy to make decisions on their own. The laissez faire manager therefore adopts a hands-off approach to overseeing his team members. In large organizations, exceptionally, the project manager may pay closer attention only to the top 10% and bottom 10% of his team members. More challenging tasks may be assigned to team members in the top 10% echelon to provide them with more fulfilling work experience. The bottom 10% needs to be counselled and in some organizations, contracts of services have to be terminated for recalcitrant cases. Management styles generally reflect two concerns; namely concern for people and concern for production. It is not desirable to overly emphasize one concern to the detriment of the other. The ideal situation is for the project manager to place strong and equal emphasis for both people and production.

9.5 Roles and Responsibilities

a. Clear definition is necessary

People are hired in organizations to assume different roles and responsibilities. It is therefore necessary to clearly define what their roles and responsibilities are to avoid misunderstanding. Similarly, the roles and responsibilities of project team members should be made known to other team members so that everyone knows exactly who is responsible for what. Defining the role of an individual in an organization is crucial because this identifies the person accountable for a specific sphere of work by a title. For example, the job title of a quality manager identifies him as being responsible for all quality matters within the project organization. Likewise for the person with the job title of safety manager who is responsible for all safety matters in the project organization. Being given a role or job title means that the person concerned carries with him the responsibility for completing the work assigned to him in the related area. Roles and responsibilities are therefore intertwined.

The person given the role and responsibilities must also be granted the authority to make decisions and the power to use organizational resources. The person granted the authority and power should also have the competence and skillsets necessary for him to take on the role to discharge his responsibilities to complete the tasks assigned. For example, there are different grades of registration for people undertaking electrical works. These grades determine who has the competence to be responsible for electrical works that involve different degree of difficulties. The different grades are the professional engineer, licensed electrician and electrician. There may also be situations when the market is booming where demand outstrips supply for a particular skillset. For example, tower crane operators may be in short supply during a construction boom.

9.6 Organization Charts

a. What the organization charts show

A formal structure must be built into an organization so that everyone knows who the boss is, who reports to whom and where the buck stops. This is facilitated using an organization chart which shows the relationships between team members, their lines of reporting and the protocol for communication flow within the organization. The organization chart may take on different forms. The most common form is one that is based on functional departments which is usually practiced in head offices of companies. Each project should also have an organization chart to show the relationships between team members and their lines of reporting. There are also more complex organization charts such as the matrix organization chart which may combine the functional structure with the project structure. When an appropriate form for the organization chart has been determined, it is necessary to source for people to fill the roles identified in the organization chart. This is where the existing staff pool is examined to evaluate if the people needed for the project are already available within the organization and are ready to be deployed.

Apart from availability, the evaluation should also consider their relevant experience, abilities, willingness and interests as well as associated costs. In the case of a large organization concurrently running a few projects, the project team members for a new project are likely to be pre-assigned from a recently or soon to be completed project. In the event where it is not possible to synchronize the start and completion time of projects to facilitate pre-assignment, the project manager will need to negotiate with other projects to share the needed resources among these projects. For example, if the on-going projects are small projects, it may not be feasible to have a full-time safety manager at each project site. Instead, a roving safety manager will divide his time between these small project sites as well as the new project if it is also a small project. Nevertheless, if the existing staff pool in the organization is unsuitable for the new project, there is no alternative but for the project manager to source for team members from elsewhere. The recruitment exercise is then activated. Alternatively, another option is for the project manager to outsource portions of the project to subcontractors.

9.7 Assembling the Project Team

a. Project team directory

It may not always be the case that everyone in a new project team knows one another. The new project may be the first time that they are working together. For this reason, it is necessary to create a project team directory after all members of the team have been assembled. Building on the project organization chart, the project team directory shows the names of all team members, their telephone numbers and email addresses. If the project team consists of members who are stationed overseas, their contact details, internet addresses of websites as well as mailing addresses should be included in the directory. It is a useful practice to include recent colour photographs of all team members in the directory so that people using the directory can put the name to a face. This is especially helpful if the project team directory is also given to the key stakeholders such as the clients and vendors.

b. Developing and managing team members

The organization chart and team directory identify the team members who are working in the project. The information in the organization chart and team directory is shared with all project team members alongside information relating to their assignments, roles, skillsets and abilities to complete their respective tasks. This also marks the start of the process for identifying their developmental needs. The organization chart and team directory should be read in conjunction with the project plan. Operationally, these documents highlight how the team members communicate with one another to perform their functions. In addition, the human resource management plan shows the timings when team members will come on board the project to do their respective parts and when they leave. For example, the human resource management plan should identify how many bricklayers are needed for the job, when they arrive at the project site and when they complete and leave.

At the same time, reports on their performance will be prepared to assess the quality standard achieved and to report if the work is completed in a timely manner. While performance reporting is generally conducted based on in-house information, it may also be necessary to solicit external feedback on performance from other stakeholders at the organizational interfaces. Timely actions should then be taken by the project manager to address any outstanding issues or problems identified in the performance reports. The measures taken by the project manager is contingent on him wielding the necessary power within the organization. There are generally five types of power which the project manager can rely on to develop his project team. The first type is expert power where the expertise possessed by the project manager enables him to lead the project team members. The second type is reward power where the project manager is in a position to recommend project team members for rewards. Such rewards may be monetary or non-monetary in nature such as bonuses and extra off-days respectively.

The third type is formal power which is related to the position or designation of an individual. For example, the designation of the project manager is a formal position within the organization that gives him certain rights and authority to issue instructions to his subordinates. The fourth type is coercive power which if invoked is likely to penalize the wrongdoer. For example, the project manager possesses coercive power to hold back payments to vendors if they produce inferior quality works. Likewise, a recalcitrant employee may not be recommended for a wage increase. Finally, the fifth type is referent power where team members know the project manager personally because the chief executive officer introduced him as the project manager to lead the project. The project manager then draws referent power from the chief executive officer's introduction. While there is a suite of power types which the project manager can rely on, wielding of such power should be carried out tactfully without adversely affecting the overall project team morale.

The project manager should actively build up his team members through bonding and developmental activities. This is important for the project because good team-work leads to better productivity. There are many ways to develop the team. For example, appropriate training can be provided to team members to help them improve their competencies. Getting team members involved in planning and decision-making can help to obtain their buy-in. They are more likely to be dedicated to the tasks which they have a hand in planning. Team bonding activities can also take place outside of the project site. These can include events such as family days, picnics and sports events. Apart from enhancing team relationships, such activities also provide opportunities for them to learn about interpersonal skills. In the midst of managing a project, there may also be occasions where disagreements and disputes occur among team members. In anticipation of these happening, the project manager should set the ground rules on how such disagreements are to be handled.

To do so, it is also necessary for the project manager to possess general management skills to lead, communicate, influence, negotiate and solve problems to get the job done. A transparent system of rewards and recognition should also be put in place to incentivize project team members for doing their jobs well. The reward and recognition system should be formally made known to all project team members and the targets must be achievable. It should also be recognized that projects are becoming increasingly international in outlook. Hence, the project manager should be sensitive to the influence of different cultures on the reward system. In some cultures, it may be inappropriate to reward an individual at the expense of the group. The reverse may also be true in other cultures. Fundamentally, the project manager should be sensitive to how cultural differences, customs, traditions and practices affect rewards.

This sensitivity should be reflected appropriately so that the rewards given will not offend anyone. In assessing the performance of team members for the purpose of recommending rewards, the project manager should focus his attention on several issues. These include the degree of difficulty of the work, the results, attitudes of the individuals, achievements of the entire project team and if there are any interpersonal conflicts between team members. To obtain a complete picture for the performance appraisal of the project team, the project manager can consider using the 360 degree approach for this purpose. The 360 degree performance evaluation will include feedback from peers, supervisors, managers as well as subordinates of the project team members. This provides the project manager with a helicopter view of the performance of an individual in the project team.

9.8   Managing Disagreements and Conflicts

a. Why disagreements and conflicts happen?

Organizations are made up of people who bring along with them different values, beliefs and practices that may clash with those of others'. Firstly, conflicts primarily happen when there are demands on limited resources and people compete to use these resources for their own tasks. For example, a project site has only one tower crane with several subcontractors waiting to use the crane. Apart from limited resources, there are also other reasons for conflicts. The second reason for conflicts relates to scheduling differences where parties are not able to synthesize their respective operations to meet a common deadline. The third reason for conflicts relates to differences in priorities among the parties. For example, in a large condominium project, the developer may indicate his priority for one block of apartments to be completed first for the show-flat to be located in that block and for marketing to commence. Differences in priorities can in turn cause scheduling disagreements.

The fourth reason for conflicts relates to different technical beliefs where parties have different approaches to the work concerned. For example, in an international project, a British engineer may disagree with the technical specifications provided by a Japanese engineer. The fifth reason for conflicts relate to administrative policies and procedures that may be obscure, overly bureaucratic or outdated. For example, because of unclear instructions, the wrong set of "permit to work" application forms was submitted for approval. The sixth reason for conflicts relates to project costs that have increased beyond the reasonable estimates. For example, the increase in costs may occur because of poor productivity or sudden increase in material costs. Finally, the seventh reason for conflicts relates to clashes of personalities. This can happen when people join an organization bringing with them their past baggage. They may have worked together in a past project with unpleasant experiences. Such unpleasant experiences may lead to personality clashes if left unresolved. It is recognized that some reasons are more common in causing conflicts compared to others. Competition for limited resources appears to be the most common reason for conflicts to surface in projects.

b. Resolving conflicts

It is in the project manager's interests to resolve all conflicts as soon as they happen. If left unattended, the conflicts may fester and spiral out of control to cause irreparable damage to team relationships as well as undermine project success. There are fundamentally five ways to resolve conflicts. The first way to resolve conflicts is through problem-solving. When a problem arises, the parties face the problem head-on and study the possible solutions that they can take to resolve it. Essentially, parties will try to find the best solution to the problem. This process is however possible only if there is sufficient time for the parties to study the problem and address the various issues faced before coming to a consensus on the best course of action to take. This approach helps to build trust and relationship between the parties and represents a win-win situation for everyone.

The second way to resolve conflicts is through forcing. This approach depends very much on the dynamics of power-play, rank and seniority between the parties. When faced with a problem and time is of the essence, the party with the power makes the decision which may not necessarily be the best decision. However, while bull-dozing through to resolve conflicts is fast, it does not help to build trust and relationship. Instead, it undermines project team development and is a win-lose situation. Nevertheless, the forcing approach to resolving conflicts is appropriate when the stakes are high, decisions need to be made quickly and relationships are not important.

The third way to resolve conflicts is to settle for a compromise between the parties. When a problem surfaces, both parties suggest their respective solutions for consideration. Instead of settling exclusively for one or the other solution, both parties adopt a shared or blended solution by taking some parts of each from the two suggested solutions. In the process, both parties give up something. Party neither loses nor wins. But this is nevertheless a "lose-lose" situation because going down this route may not yield the best solution. Parties compromise on a solution because they have an equal relationship and they do not wish to pick up a fight.

The fourth way to resolve conflicts is through smoothing the problem. In this approach, the parties smooth out the perceived size of the problem by minimizing it. Even if the problem is large, both parties will pave a way for this to be seen as not really a big problem. This offers a temporary respite from the problem that allows parties to calm themselves in order to avoid aggressive arguments. Rendering a large problem to a perceived small problem through smoothing is however a "lose-lose" situation because no one party actually wins in the long run. Smoothing is however acceptable when time is of the essence, when no one is able to propose a solution that is workable and when the issue at hand is not crucial. It is also a useful approach to maintain relationships among the parties.

The fifth way which some parties adopt to resolve conflicts is simply to walk away from the problem. Walking away or withdrawal offers an immediate solution but is one of the worst conflict resolutions possible. This is especially so when one party walks away angrily in disgust. Nevertheless, withdrawal from the problem provides the parties with a cooling off period that avoids damaging confrontation. This approach is acceptable if the issue at hand is not critical. The project manager should also appreciate that some conflict resolution approaches work well in some cultures but not in others. For example, in the Confucian society where respect for seniority is important, parties will generally accept the views of a senior person in resolving conflicts. Finally, managing people is also closely integrated with project communications management. Otherwise, how would they know unless the project manager tells them?

9.9 Revision Questions

1. Why are people the most important asset in project organizations?
2. Why is dealing with human resource management of a multi-faceted nature?
3. What constitutes human resource planning?
4. What are organizational interfaces?
5. What are technical interfaces?
6. What are interpersonal interfaces?
7. What are logistical interfaces?
8. What are political interfaces?
9. How do the project teams identify staffing requirements?
10. What are some of the project constraints affecting human resource planning?
11. Can the project manager use past project experiences to complete organizational planning?
12. What are some of the good human resource management practices for staffing a project?
13. Why is it necessary for the project manager to understand various organizational theories?
14. What lessons can the project manager draw from Maslow's Hierarchy of Needs?

15. What lessons can the project manager draw from Herzberg's Theory of Motivation?
16. What lessons can the project manager draw from McGregor's Theory X and Theory Y?
17. What lessons can the project manager draw from Ouchi's Theory Z?
18. What lessons can the project manager draw from Vroom's Expectancy Theory?
19. Which is a better management style: autocratic or democratic management style?
20. What are the linkages between roles, responsibilities, power and authority?
21. Why is an organization chart important for project management?
22. What should the project manager consider when examining the staffing pool?
23. What key information should the project manager collate when assembling the project team?
24. What are some of the considerations for developing the project team?
25. What are the five types of power that managers may possibly possess?
26. What are some useful team-building activities for the project team?
27. How should project team members be rewarded?
28. Why is performance appraisal needed for project team members?
29. What are some of the reasons for conflicts to happen in organizations?
30. What is conflict resolution?
31. What are some of the approaches that may be adopted to resolve conflicts?
32. Is it necessary for the project manager to also function as the leader to pre-empt conflicts?
33. Why is human resource management closely aligned with project communications?

Chapter 10
Project Communications Management

10.1 Introduction

a. Need for communications

People need information and instructions in organizations. Communications is the life-line of projects as it provides the linkages between management, clients, project team members and other stakeholders. This life-line focuses on who needs what information in what format and when. It is to be expected that larger projects need more information which also heighten planning for communications. First and foremost, information is pervasive and permeates at all levels and in all aspects of organizational life. How would one know what the project scope is unless it is communicated to the stakeholders? Similarly, information relating to resource requirements is necessary for the project team to procure these resources in time for use. Communications is therefore needed to convey resource requirements as well as schedule deliveries to the project. Without clear, detailed and timely communications, there would be confusion.

Incomplete information may be given that cause project stakeholders to do the wrong things. In the worst case scenario, the project may even grind to a halt for lack of information because communications channels have failed. Communications is however not confined only to stakeholders who are directly involved with the project. In a school refurbishment project, for example, the stakeholders are various members of the design and construction teams, school management committee members, principal, teachers, students, parents, alumni, canteen operators, vendors, etc. Because the coverage for communications can be wide in projects, different stakeholders may have different expectations concerning the information that they receive, in the format they want, and when. A clear balance needs to be reached in providing sufficient information in a timely manner and in an appropriate mode.

© Springer Nature Singapore Pte Ltd. 2018
L.S. Pheng, *Project Management for the Built Environment,*
Management in the Built Environment,
https://doi.org/10.1007/978-981-10-6992-5_10

Providing too much information leads to information overload, contrasting sharply with an extreme case where no information is provided at all. There are generally five areas of responsibilities where communications in general is concerned. Firstly, internal communications formulate and carry out effective communications with employees. This can take the form of providing sufficient communications tools to employees as well as setting out the protocols and procedures for internal communications. Information relating to employee welfare, human resource matters, training opportunities, etc. should be made readily available to employees as part of internal communications. Secondly, external communications make use of various communications channels to position the organization in a strategic position to the outside world. This may take the form of advertisements in newspapers, magazines and television. At strategic locations in busy public places such as train stations, bus interchanges and expressways, bill boards can also be used as a form of external communications by the company to publicize its products or services.

However, to achieve strategic outcome for the company, external communications should be short and meaningful, using memorable taglines or themes for this purpose. For example, Mercedes Benz adopts the slogan "The best or nothing" for this purpose. Thirdly and closely aligned with external communications is media communications. This is where a targeted approach to communications is made to cultivate and build good media relations with a view to avail the company of positive review coverage especially in the areas of financial performance and international outreach. This requires the company to be in close touch with print and television journalists for press releases to be made available to them for reporting.

Fourthly, investor relations require effective communications to provide timely and useful information to the larger financial community as well as individual investors. This can take the forms of regular financial updates and reports as well as annual general meetings. Fifthly, community relations require constant communications with the government, non-governmental organizations and surrounding communities to develop positive relationships that will eventually brand the company as a socially responsible organization. For example, through sponsorship of a neighbourhood sports event, the company is able to play its part to create awareness of a healthy lifestyle in the community. These five areas of corporate communications responsibilities when viewed in the context of project communications management highlight the extent of coverage of communications in projects.

b. What project communications management entails

Being able to communicate clearly and effectively is one of the most important skillsets of a project manager. Communications in particular is very important in the context of human resource management. To be able to manage people, one needs to communicate. In the workplace, a significant portion of the time spent by employees is on communications. This can be in the form of face-to-face meetings, telephone calls, email messages, reports and presentations, etc. Project communications management therefore plays a key role in organizations. It focuses on what information is needed by who, when it is needed and in what format. Having understood the information needs, a project communications plan is then produced to provide the information needed. The plan includes the generation, collection, dissemination and storage of information.

Information is critical to keep all stakeholders in the know so that everyone is clear about how the project is progressing. This is especially important in for example upgrading works in public housing programs where it is necessary to maintain clear and open communications with the residents. Because such upgrading works cause dust and noise, communications help to update residents of progress. This is to minimize the inconvenience caused to them during the upgrading project. Successful and effective communications provide the key link between stakeholders by conveying ideas and information between them. In essence, how would they know if you do not tell them? Project communications management encompasses the five processes of meeting, planning for communications, distributing information, reporting performance and stakeholder management. These five processes align closely with the project management processes of initiating, planning, executing, monitoring and controlling as well as closing respectively.

10.2   Communications Planning

a. Factors to consider

To initiate project communications, it is necessary for the relevant stakeholders to meet to plan. Communications planning essentially focuses on who needs what information, when the information is needed, and in what format. This is completed at the early stage of a project but with provisions made for updates in the future through action items indicated in minutes of regular site meetings. A range of information types is involved in communications planning. For example, these can include the quantities of concrete required that are communicated to the vendors for timely deliveries or the colour of the carpet in the common room selected by teachers in a school building project. As the project progresses, the needs of stakeholders may change.

Consequently, in the course of the project, stakeholder needs, information types requested for, formats in which the information is disseminated, etc. should be reviewed, checked for accuracy and updated for records. The formats in which information is disseminated can use different modalities such as drawings, brochures, architectural models, meetings, email messages and telephone calls as the situation calls for. In the construction industry, project information types can be classified at three distinct levels. Firstly, at the high level, information types can include information relating to plot ratios, zoning requirements, legislative provisions, etc. At the mid-level, information types can include information relating to structural details and mechanical and electrical systems for the building. At the detailed level, information relating to choices of paint colours, textures of tile finishes, etc. is provided for activities to take place. Hence, there is a spectrum of information types in projects at the high level, mid-level and detailed level.

Beyond this classification, the manner in which information is conveyed is just as important as what is being disseminated. There are many factors that can affect communications planning. Organizational culture and structure do play a part in influencing how information is disseminated and the responses one is likely to receive from the other party. Communications need to consider national cultures and their influence on the modalities in which the information is sent. For example, transmitting information between Japanese stakeholders and American stakeholders may need to account for seniority and hierarchical constraints. Expectations relating to formal or informal modes of communications have to be considered in the plan to accommodate cultural and structural differences in organizations.

There may also be industry-specific regulations and standards which the communications plans need to account for in the project. This includes for example the need to digitalize the information before on-line submission as this is a regulatory requirement in that industry. Communications planning also considers if the information communications and technology infrastructure is in place in the project organizations to facilitate ready dissemination. For example, in a rural project, the availability of in-house servers and internet access should be considered. Likewise, the availability of appropriate human resources to support information generation, collection, dissemination and storage should be considered. Hence, for example, in projects that may generate much public interests communications planning should consider the availability of marketing communications professionals who can take on the role of creating a positive image for the company. Marketplace conditions can also affect communications planning. For example, when the market is quiet, consultants with less jobs on their plates can afford to spend more time on project communications.

Similarly, if there is intense competition between consultants, the likelihood of low consultancy fees is high. As a result of the low fees received, the consultants may be reluctant to spend more time to provide details to the project team. The extent to which communications planning takes place can also be influenced by the risk tolerances of stakeholders. When risk tolerance increases, the provision of less information may be accepted and tolerated, as a result of which the risk of mis-interpretation may increase between stakeholders. For effective communications planning to take place, it is worthwhile for the project team to consider implementing a real-time project management information system. Such a system facilitates timely information sharing to correspondingly support project integration management.

b. Identifying communications demands

Historical information and lessons learned from past projects are useful starting points to identify communications demands in the current project. It is also useful to refer to the project scope statement to determine what should be included in communications planning. Bridging the project scope statement with communications planning is important for the project team to manage stakeholder expectations as the scope statement is a reference point for the project to be anchored. Nevertheless, in the process of identifying communications demands, the constraints and barriers faced should also be recognized. Such constraints include different locations and time zones within which stakeholders are in, language barriers and availability of tele and video-conferencing facilities. Where necessary, there may be a need for translation services to be made available during discussions.

Other constraints may include different technical capabilities, incompatibility of communications software and subscription to different codes and standards in different countries. Additional resources may be needed to overcome these constraints and barriers such as regularizing different technical competencies to a common benchmark, engaging the services of an interpreter and calibrating different communications equipment for compatibility. Identifying communications demands may also be influenced by the assumptions made. Such assumptions should be clarified at the first available instance. For example, it should not be assumed that top management requires project reports to be sent by the project manager through email messages. Checks should be made with top management on the format as well as frequency in which performance reporting is to be conducted for them.

If the preference is for official progress reports as well as daily update memos to be made to top management, this preference should be incorporated as part of the communications demands for the project. Hence, such clarification should be made clear right from the start because there is no point in spending much time and efforts to generate voluminous reports which no one is interested in reading. In identifying communications demands, it is also worthwhile to adopt a one-page executive report or summary approach for top management's information. A key tool in identifying communications demands is the organization chart at the head office which shows the hierarchy of who reports to who, and through whom.

For projects, the project structure indicates a similar hierarchy albeit at the level of the performing organization. The project structure can be extended beyond project management to program management, portfolio management and project management office, each with its own communications requirements. Communications demands also depend on the responsibility relationships of stakeholders. For example, in a matrix structure, the project team member reports to two superiors in the project structure as well as the functional structure. Identifying communications demands therefore need to consider the different functional departments and professional disciplines who are involved at various times in a project. For example, an employee from the information technology department is involved with setting up the computing facilities at the early stage of a project. He then leaves the project only to return whenever there is a need to troubleshoot intermittent problems and at the end of the project to dismantle the computing facilities.

Communications demands also depend on the number of project stakeholders involved and where they are located. For example, in retrofitting an old Chinese temple in Singapore, there may be a lack of skilled artisans locally who can repair intricate wood carvings. Hence, such skilled artisans from Hubei, China need to be engaged for this purpose. This creates communications needs between the project team based in Singapore and the contracting firm in China who employs these skilled artisans. It is also clear that communications demands are derived both externally and internally to the project organization. For example, internal information needs may be requested by the board of directors while external information needs may be directed to the media. Equally important is information provided to stakeholders or members of the public whose daily routines are affected by the project. For example, in upgrading the public amenities in a housing estate, parts of the estate may be barricaded for the safe erection of sheltered walkways. Residents living in the estate are therefore inconvenienced to some extent. Timely information should be provided to the residents to update them on progress as well as when the project is likely to be completed.

The greater the number of stakeholders affected by the project, the greater is the need for more communications channels to be created. The number of communications channels can be calculated using the formula $N(N - 1) \div 2$ where N is the number of stakeholders identified. Hence, if there are 5 stakeholders in a project, the potential number of communications channels is $5(5 - 1) \div 2 = 10$. As more stakeholders are involved in a project, the number of communications channels increases which also correspondingly increase the likelihood of information loss and misinterpretation. This however depends on the length and details of the piece of information and how it is conveyed.

10.3 Communications Plan

a. Factors affecting the plan

Depending on the size and complexity of the project, many factors can surface to affect the project communications plan. One such factors is the urgency in which the information is required and for whom. For example, if there is an urgent request from the chairman of the board of directors requesting for information to respond to a query from a regulatory authority, that information needs to be generated and transmitted quickly. Information in response to emergencies such as a serious site accident should however be prepared beforehand as part of business continuity management. The availability of compatible information and communications technology, intranets and project management information systems that is well staffed should help to ease communications planning significantly.

This is especially so for projects that requires a long period for completion as such projects are also likely to be more complex. Another factor that can affect communications planning relates to the project environment where on the one hand, there is a project team in the same physical location versus another virtual project team that is located overseas. Geographically speaking, communications clarity becomes less as physical distance increases. Apart from communications technology, there are also many different ways in which communications can take place. These can include informal corridor meetings and formal project meetings to update on progress status. Face-to-face informal meetings can also take place in the canteen, thus eliminating the need for long and time-consuming discussions over email messages.

Such informal meetings should however be followed up by written notes of confirmation that can be tracked using on-line databases. When communicating, the level of security relating to the piece of information should also be assessed. Information should be classified as private, sensitive, confidential or secret and with the appropriate measures taken to prevent breaching privacy laws and information leakage. For example, media embargo until the official release date should be complied with.

10.4 Communications Management Plan

a. Contents of the plan

The contents of the communications management plan are both diverse and comprehensive. The plan first identifies who the project stakeholders are. Stakeholders' identities are necessary for the project team to know who they are in order to understand their communications requirements. This is to ensure that only the information that is appropriate is provided to them. For example, if ready-mixed concrete will be used in the building project, then the ready-mixed concrete vendor is a stakeholder. The vendor needs to know the quantities of concrete to be delivered to the project site and when deliveries should be made. The communications management plan also provides information on the modes to be used to distribute or gather information depending on the nature of the tasks. For example, the submission of formal progress status reports for top management information as against impromptu email messages to update project team members on work progress.

The communications management plan includes instructions on how information is to flow from the project to the correct recipients. This is to ensure that information does not fall into the wrong hands. For high security projects, sensitive information may need to be encrypted before distribution by emails. An added level of security may be added where confidential files are to be opened only with passwords. The plan may also include instructions relating to who such email messages should be sent to for approval before distributing. The appropriate methods of communications for different types of recipients should also be set out in advance in the plan. For example, if the company faces a crisis and needs to deal with the media, the communications management plan should provide instructions on the appropriate communications method to use and identify the person responsible for the task. In planning ahead, the communications management plan has a schedule of when different types of communications are likely to occur.

These can be categorized as regular communications or ad hoc communications. The former may include the provision of information for monthly progress payments or regular reporting to the board of directors. The latter may include press releases or one-off enquiries from schools requesting for permissions to visit the project. However, not all scenarios can be accounted for even with the most detailed planning in place. In exceptional cases, when issues at a lower level cannot be resolved within a given time frame, such issues should be escalated to higher management for a solution. For example, in a school redevelopment project, a well-to-do alumnus may decide to donate a substantial sum of money mid-way through the project to build a swimming pool. The pool was not originally part of the project scope and a decision for its addition is outside the authority of the project manager. In this case, the generous donation to add a swimming pool in the school redevelopment project should be escalated to the higher management of the school to decide.

To ensure consistency and the retrieval of information readily, the communications management plan should specify the methods for information classification, coding and storage. Standard coding systems are preferred as these are more widely recognized and understood by more people. The use of a coding system for information storage would also facilitate easy retrieval for updating information as the project progresses. The communications management plan provides instructions on who is responsible for and when information should be updated at specific milestones. For example, if there are plans for a roof-topping out ceremony by an important guest, the date of the ceremony is a key milestone for the project. If there are project delays and the ceremony needs to be postponed, timely information to this effect should be distributed to the people concerned.

Finally, if the project is a complex one involving the use of new technologies and procurement modes, the communications management plan should include a glossary of terms and definitions for explanation. This serves to avoid misunderstanding among the project stakeholders. For example, if the project includes the creation of clean rooms for the production of semi-conductor peripherals, the glossary of terms should provide a clear explanation of what a clean room is all about.

b. Preparing to distribute information

Preparation for information distribution is closely linked to human resource management where stakeholders within the project organization are first identified. This is then extended to identifying stakeholders outside of the project organization such as vendors and government agencies. Preparation for information distribution ensures that only the appropriate information is sent to the correct stakeholders at a time when it is needed using the proper distribution modes. The preparation process is fundamentally about the execution of the communications management plan. Three inputs trigger the need for information distribution. Firstly, the results of performance reporting may either be good or bad. For example, is the waterproofing membrane in the bathroom area laid correctly? If it is laid correctly, then information to this effect is distributed to the relevant stakeholders for acceptance. If the work results from laying the waterproofing membrane are not satisfactory, likewise, information to this effect is distributed to the relevant stakeholders for rectification work to commence.

Secondly, implementation of the communications management plan also triggers information distribution. Essentially, the communications management plan uses the work breakdown structure as a starting point to demarcate the various stages of a project. The plan is then developed to include details on the type of information required, deadlines of when, where and how the information is to be obtained, and who the information is to be distributed to. Thirdly, the project plan is also a trigger for information distribution. The project plan provides descriptions of information requirements as well as conditions that are integrated with communications. The project plan serves as a guide on for example what project information is needed next week. Such information may be as simple as the colour of the wall tiles chosen by the client.

10.5 Skills for Good Communications

a. Setting up communications

Good communications can be affected by many factors. These include the individuals involved, namely the sender and the receiver of the information. Quite apart from face-to-face communications, information distribution often requires an encoder and a decoder such as the sender's mobile phone and the receiver's land-line telephone system respectively. Connecting the encoder and the decoder will be the medium, which in this case will be the phone lines in the land-line telephone system. The transmission of information can be affected by noise such as poor connection or loud music in the background. If the communications is successful, acknowledgements should be given by both the sender and the receiver to this effect. Acknowledgements are important signals to indicate that communications is clear and that the information has been sent, received and understood.

b. Communicating successfully

Apart from noise, there are five other factors that can influence successful outcome of the communications process. Firstly, the para-lingual factor relates to the tone and pitch of the sender's voice that can affect the message being sent. The tone may be monotonous and unexciting to cause the receiver to lose interest in the communications process. Secondly, the feedback provided by the receiver at the request of the sender allows the latter to confirm that the former has indeed understood the message. Feedback can be triggered by the sender asking for a response from the receiver or by encouraging the receiver to ask questions for further clarification. In the communications process, written feedback is useful as this provides documentary records for future reference.

Thirdly, active listening by both the sender and the receiver is a reflection of basic courtesy. Through active listening, both parties are able to confirm the message by asking questions and prompting for clarity. This is especially useful in brainstorming sessions to identify creative solutions to problems. Fourthly and closely aligned with active listening is effective listening. In effective listening, the receiver is engaged in asking relevant questions that shows up the effectiveness of the communications process. If the questions asked are not pertinent to the issues discussed, then effective listening has not taken place.

Lastly, the communications process may also be distracted by nonverbal cues. These can take the form of body languages such as intense frowning, gesticulating hand gestures or an expressionless face that looks lost. Appropriate responses have to be made to deal with specific body language in the communications process. For example, if there is a facial expression that looks lost, then the message needs to be repeated with further clarification made. Some body languages may be intimidating such as when the tone is aggressively forceful. This is where the communications process needs to be moderated for a win-win situation to evolve. Stakeholders should appreciate that how something is said is just as important as what is being said. Face-to-face conversation is always beneficial as it allows the stakeholders to observe the body language and the tone while listening to the message at the same time. It also offers stakeholders an immediate opportunity to seek clarification spontaneously to avoid misunderstanding.

While direct face-to-face conversation is the preferred mode for communications, this may not be possible all the time. Hence, stakeholders have to resort to using various non-direct modes of communications such as written reports, letters, memos and email messages. In the internet age, email messages are by far the most prevalent mode of communications in projects. However, there are several downsides to using email messages. Email messages can be quite impersonal wherein the tone and body language of the project stakeholders cannot be observed. When an individual is inundated by many email messages daily, there is less time to frame proper replies as a result of which disputes may happen. Because replying to email messages is so easy, individuals may simply hit the send button after crafting an angry reply. This should be avoided by consciously waiting out a cooling down period.

10.6   Information Retrieval Systems

a. Types of retrieval systems

Information distribution starts and ends with the information retrieval system. For distribution to commence, information is first retrieved from the information retrieval system. When information distribution is completed, the responses from other stakeholders are stored in the information retrieval system. The types of information retrieval systems to use depend on the size and complexity of the projects. For simple projects, information retrieval systems can make do with a simple manual filing system using a lockable metal cabinet. However, for large projects where demands on information are fast-moving and complex, advanced information storage databases can be used. These can take the form of a repository supported by an integrated information system developed in-house or a suite of robust project management software that can readily be bought off the shelf.

Information retrieval systems should facilitate ready retrieval of the correct information which is organized by a recognizable coding system that is readily accessible and secured. The information retrieval system should not lead to information overload by requiring the user to comb through large volume of information before finding the correct one. The system should be easy to navigate and update for information storage as well as retrieval. This should be a key feature of the information retrieval system because information is continuously required throughout the entire duration of the project. Apart from the written format, information can also come in vastly different forms ranging from drawings, mock-ups, architectural and virtual models. Information if in the written format may also be expressed in different language. Hence, the information retrieval systems should be able to account for the wide range of information requirements and specifications.

Essentially, the best format to present the information is the one that is most appropriate to convey the information. Because information can be conveyed using different modalities, this also influences the corresponding distribution methods. Hence, there is no one-size-fit-all approach to distributing information. Consequently, information can be distributed through project meetings, hardcopy documents, facsimile messages, email messages, telephone calls, videoconferencing, shared databases or a dedicated project website.

b. What happen after information distribution?

Information retrieved and distributed to stakeholders will return eventually to the sender for storage. This is after the receiver has provided a response based on the information received. There are several results after information distribution is complete. Firstly, there may be lessons learned from the correspondences that now form part of the historical records of the project organization. Project records that track contractual obligations are also stored. These records may for example include signed agreements and payments made. Similarly, project performance reports such as those relating to quality control checks should also be stored. When the project team communicates with stakeholders through presentations with the objective of receiving feedback from them, such presentation materials and stakeholder feedback should also be tracked and stored.

The presentation may relate to a raised floor system recommended for concealing cables in computer rooms for the purpose of soliciting feedback from the client. As the project progresses, stakeholders may need to be notified of outstanding issues such as those relating to change requests, their approvals and overall progress of the project. Such notification is served through information distribution.

For example, if a public housing estate is being upgraded, the residents need to be kept informed of work progress or delays on a regular basis. Such information should be distributed in a timely manner. The project team may make use of notice-boards at the building site for information distribution. The project team may also set up a dedicated Facebook page or other appropriate social media platforms to disseminate information as well as provide an avenue for residents to provide feedback and suggestions.

10.7   Performance Reporting

a. Purpose of project performance reports

This process facilitates monitoring and controlling to ensure that everything is progressing as planned in the project. The performance reporting process essentially collects, organises, analyses and disseminates information on how resources used are contributing to the project objectives. The analysis may be focused on different objectives such as those relating to costs, time, risks and quality. For example, if hazards and their corresponding risks have been identified in the project, performance reporting would highlight if the persons responsible for implementing the mitigation measures have completed their tasks successfully. Performance reporting is important because the stakeholders paying for the resources need to have the assurance and confirmation that the desired outcomes are achieved and that things are progressing as planned in the project. The client may therefore request for regular meetings with the project team to review performance. Such meetings further open up communications channels between the project stakeholders.

b. Analysis of project variances

During the course of the project, variances affecting scope, time, quality, costs and resources may occur. It is therefore necessary to identify such variances when these occur with a view to take preventive or corrective action. Analyzing project variances is an ongoing task as the project progresses. The analysis serves to establish the root causes of the variances and the corresponding actions needed to prevent these from happening again in the future. The causes can be varied and range from anomalies or simply flawed estimates on budget and schedule. The results from analyzing project variances should be communicated to the stakeholders affected. The project stakeholders should then further establish if the variances are within the acceptable range; for example, a costs variance between −5 and +10% may be considered to be acceptable. However, this can depend on the stage the project is in. At the early stages of a project, variances are expected to be larger due to lack of detailed information available for use. As the project moves closer to completion, variances are expected to be lesser because any late-stage variances are likely to be more expensive and disruptive for the project.

A positive outcome from the analysis of variances is that the project team can utilize the findings to determine if similar variances are likely to occur in the near future. Similar to analysis of variances, results of performance reporting should also be communicated to the relevant project stakeholders. For example, performance reporting may be based on Earned Value Management where the amount spent is compared with the monetary returns. Hence, the outcomes of all performance reporting will lead to performance reports covering various aspects of a project such as quality compliance and scheduling progress. Reports are channels of communications with project stakeholders. Performance reports will highlight if the project is likely to complete in time and within budget. If performance reporting is not positive, corrective or preventive measures should be taken immediately to avoid further deterioration of the situation. It may also be the case that the measures taken in response to performance reporting may in turn lead to change requests and recommended corrective actions.

10.8   Stakeholder Management

a. What this entails

Communications is about getting the right information to stakeholders to obtain their buy-in. Providing the necessary information to stakeholders is also a form of closing where project communications management is concerned. The project manager should therefore engage stakeholders actively and to answer all the queries that they raise. It is vital to communicate with stakeholders on a regular basis to update them on the project status and progress. Managing stakeholders effectively through clear and timely communications help the project manager to resolve stakeholder issues openly. This serves to promote project buy-in from the stakeholders which in turn provides synergy to the project to both minimize as well as limit disruptions.

Good team relationship can be facilitated through effective communications. The process leading to the cultivation of project buy-in depends to a large extent on the communications management plan. The plan helps to direct pertinent information to the correct stakeholders at the time when it is needed. Information dissemination can take various forms including email messages and telephone calls. It is however a good practice to follow up a telephone call with an email message for confirmation. Where possible, face-to-face meeting with stakeholders is the best mode for communications as this allows instantaneous questions and answers to provide clarity as well as resolve issues. Likewise, such meetings should be followed up with documenting the issues discussed in a formal document known as an Issues Log. The document sets out who the owner of the issue is and a deadline for its resolution. This is similar to the practice of having an action column in official minutes of meeting.

The actions taken to resolve the issue should also be documented formally as part of the communications records. The communications management plan identifies what type of information is needed, how the information is to be collected, organized, retrieved, distributed and stored. The plan includes a schedule of regular project status meetings which are to be expected as the project progresses. The plan also caters to ad hoc events such as when communications is triggered to deal with unexpected conditions like inclement weather and change requests.

10.9 Revision Questions

1. Why is communications one of the most important skillsets of a project manager?
2. What does project communications management focuses on?
3. Why are the four processes (namely communications planning, information distribution, performance reporting and managing stakeholders) crucial for the project manager?
4. When should communications planning be completed preferably?
5. What are some of the considerations for communications planning?
6. What are project constraints in communications planning?
7. What are some examples of project assumptions in communications planning?
8. How is the demand for communications identified in a project?
9. What are the different ways for communications to take place in projects?
10. What are some of the factors affecting the communications management plan?
11. What items does the communications management plan provide?
12. What preparation work is needed for information distribution?
13. What are good communications skills?
14. Why are the following five attributes important to create successful communications: namely para-lingual, feedback, active listening, effective listening and nonverbal cues?
15. How is the information retrieval system created?
16. What are the various modes for distributing information?
17. What are the results from successful information distribution?
18. Why is it important to trigger project performance reporting?
19. What does project variance reporting achieve?
20. What are some of the outputs from performance reporting?
21. Why is it important to manage project stakeholders as part of communications management?
22. What are some of the modes that can be adopted to manage project stakeholders in communications management?
23. Is it true that the communications management plan sets out to establish "who needs what information and when they need it"?

24. What are some of the conditions that prompt the need for meetings?
25. What three languages should I learn to be competitive as a project manager in this region?

Chapter 11
Project Risk Management

11.1 Introduction

a. What are risks?

Many people tend to associate risks with a negative mind-set. This is because risks cannot generally be anticipated as risk occurrences are typically events or conditions that are unplanned but yet can have vast positive or negative impact on project outcome. While most risks are perceived to be a threat, not all risks are actually bad. Risk is often associated with a crisis. The Chinese translation of crisis can mean a threat or opportunity depending on how one responds to the crisis. The same analogy can be extended somewhat to risks which can lead to total project failure, severe project delays, poor quality and significant cost increase. Performance reporting should indicate the extent of risks that have manifested in a project. In response, risk management is the process wherein the project team identifies the risks, analyses and ranks them and ascertains the appropriate actions to be taken to circumvent the threats. Risks can come from many different sources to affect a project. These can be political risks, economic risks, societal risks, technology risks and environmental risks.

b. Need to plan for risk management

Risk management planning is about making decisions and taking actions to manage the risks. The level of risk is however not expected to be the same for all projects. This can depend on the nature and type of projects. A simple small house-building project has less risks than a large complex cross-country civil engineering project. Similarly, painting the school fence has less risks than an exploration project drilling for oil at sea. As the priority of a project increases, risk planning should also correspondingly takes on a greater role. To safeguard employee welfare, a government agency responsible for national manpower policies may also introduce relevant legislation for this purpose. One example is the Workplace Health and

© Springer Nature Singapore Pte Ltd. 2018
L.S. Pheng, *Project Management for the Built Environment*,
Management in the Built Environment,
https://doi.org/10.1007/978-981-10-6992-5_11

Safety (Risk Management) Regulations. Under this regulatory framework, a risk manager measures, assesses, evaluates and manages the risks identified in a project site which is actually a workplace.

c. Stakeholders' risk tolerance

Projects are made up of people who have different tolerance for risks. Different risk tolerances influence how risks are perceived. Stakeholders may by nature be risk adverse and have inclinations to avoid any forms of risks at any costs. On the other hand, some stakeholders by nature are risk takers. In this context, the government agency responsible for industrial safety is certain to have zero tolerance for accidents and therefore practices low tolerance for risks. To further the safety cause, the government agency may put in place an on-line system making it a regulatory requirement for all companies to report incidents that happen on the factory floor or in a project site. The government agency may also conduct spot checks at the factory or in the project site to ensure there is no safety infringement. If such infringements are found, and depending on the severity, the agency may issue a remedial order instructing the organization to immediately carry out rectification works.

Where the safety infringement is severe or when serious accidents have already happened, the agency may issue a stop-work order that requires all works to immediately cease in the factory or project site. Because different stakeholders have different appetites for risks, their expectations should be made known at the start of the project and written down as policy statements. The different tolerances to risks may also be reflected through the different actions taken by the stakeholders during the project. The approach to managing risks is actually based on good common sense. For example, in the case of hospitals where life and death can be at stake, no shortcuts or slipshod quick fixes are allowed when installing new medical equipment in the operating theatre. On the other hand, a landscape project to create a new community garden carries little or no such risks.

However, the same cannot be said of a painter working on a gondola suspended from a high-rise building. In the painter's case, checks must be made of the worthiness of the gondola as well as his personal protection equipment such as safety harness and helmet. Risk tolerance and acceptance is clearly different between the three cases of hospital, community garden and painting work. Hence, risk tolerance describes the extent to which a person is willing to accept risks. This tolerance to accept or avoid risks influences the amount of time and money that the person is willing to reduce or eliminate the chance of failure occurring. The lower his tolerance for risks, the higher will be the amount of time and money that he needs to expend to manage the risks. Correspondingly, the higher his risk tolerance, the lower will be the amount of time and money that he needs to expend for risk management.

This should however be viewed within the context of the priority accorded to the project. Here, the higher the project priority, the higher will be the costs needed for risk management and the lower will be the utility function for the project because of increased costs. The priority of a project is therefore relevant to how risk tolerance unfolds. An example of a high priority project is the opening ceremony of the Olympics Games which is beamed live to a global audience. The opening ceremony is a high priority project to the host country because of national prestige. Risk tolerance is therefore likely to be low.

11.2 Risk Management Plan

a. What does it do?

Comprehensive planning is required to ensure that all possible risks associated with a project are identified. To achieve this aim, a risk management plan is needed. The risk management plan sets out to accomplish the following tasks. First and foremost, it determines how risks will be identified. If the current project is similar in many aspects with past completed projects, risk identification may start from reviewing the historical records of these past projects kept in the organization. If the project is new, risk identification can start by visiting the project site and reading relevant documents maintained by the local town councils. If the project involves new systems or technologies, risk identification can be initiated by consulting the experts or those with prior relevant experience.

The risk management plan will also evaluate if quantitative risk analysis is required and if so, when the analysis should be completed for the project to move on. For example, where projects involve deep excavation in poor soil conditions, geotechnical survey and engineering calculations will be required to determine the risks of ground collapse associated with the excavation. Appropriate actions are taken in response to results of the quantitative analysis. Further response may come in the form of computer modelling or simulations to further test the results of the quantitative analysis. The risk management plan also establishes how qualitative analysis to assessing risks should be carried out. This may come in the form of a questionnaire where stakeholders are asked to provide their subjective inputs using a qualitative scale where 1 indicates low risk and 5 indicates high risk. The risk management plan also prescribes how risk response planning is to take place. For example, in the event of flooding in the basement excavation after heavy rains, the plan should indicate how such a risk should be mitigated in the first place, failing which what measures should immediately kick into resolve the crisis. Effective risk response planning should also ensure that the organization is able to continue its business even while facing a crisis such as an earthquake. An important component of the risk management plan is related to how risks should be monitored.

Risk monitoring can take place through site visits and inspections. For example, the risks associated with working on scaffolds include workers falling from height and workers being hit by falling objects. Daily inspections therefore facilitate the monitoring of such risks to prevent them from happening. The risk management plan associated with scaffolding risks should therefore indicate who is responsible for monitoring, what should be included in monitoring and when monitoring should be carried out. Risk monitoring needs not always be carried out manually. The monitoring process can also be automated for example using sensors to continuously monitor ground settlements during excavation works. The risk management plan is a comprehensive plan because it covers all on-going activities that are likely to take place in the life cycle of a project. For a typical building project, the risk management plan should cover all activities associated with site clearance, excavation, foundations, superstructure works, roof, architectural works, mechanical and electrical installations, etc.

b. Classification of risks

The nature, type and magnitude of risks may vary from one project to another. In addition, many possible risks may surface in projects. It is therefore useful to prepare a list of the different risks which the project team is likely to encounter. Such risks may also be categorized appropriately according to their nature. Listing and categorizing risks help the project team to adopt a systematic approach to risk identification and thereby avoid the situation of missing out on some risks due to oversight. There are generally many different ways to classify risks. One commonly adopted approach classifies risks into the Technical category, Organizational category External environment category and Project management category. Within each category, the project team can then adopt a structured approach to identifying the list of risks in that category. For example, excavation and scaffolding risks fall within the Technical category. Risks of missing documents and cash-flow problems fall within the Organizational category.

When resources are procured overseas, payments may be at risks of currency exchange fluctuations that can be minimized through hedging. Such foreign exchange risks fall within the External environment category. The risks associated with inaccurate cost estimating, poor control and scheduling fall under the Project management category. For example, a resource required for the project is planned to arrive at a certain date for installation but did not because of scheduling risks. Risk identification is therefore carried out systematically using these categories and their corresponding lists of identified risks. Classification does help with risk identification where risks are identified and documented. It then progresses to evaluate how these risks are likely to affect the project and when.

Risk identification is however an iterative process. It does not end at its first planning. This is because circumventing one risk may give rise to another risk occurring or cause another risk to disappear. Project risks are somewhat interrelated with one another and affecting one another. There can be occasions when an accident does not happen just because one risk is present. The accident happens because a series of risks is aligned to give rise to that moment of carelessness or recklessness. For example, if a project is rushed for early completion, workers may be required to work overtime. If this occurs for an extended period of time, the workers can become tired, lose their momentary concentration, trip over and cause an accident to happen. A more structured systematic and scientific approach to risk identification and management is therefore recommended.

11.3 Methods to Identify Risks

a. Background considerations

A structured and systematic approach to risk identification is to start with reviewing the project documents. These documents can include the project scope statement, specifications, drawings and method statements. Documents relating to the work breakdown structure, components and activities should be reviewed to identify risks that may possibly occur during project execution. The method statements related to all activities should be evaluated to determine if the proposed methods would give rise to risks. Method statements are description of how activities would be carried out. For example, if concrete casting is required at upper floors of a building project, the method statement for these activities should describe the equipment to be used for hoisting or pumping the concrete as well as the number of workers required in the concreting team.

When reviewing the activity and its method statement for risk identification, the underlying assumptions for the method to be used should be tested for reliability. Higher reliability means there is less risk in that activity or that the risk mitigation measure is likely to be effective. Project team members should also spend time brainstorming over new methods, systems or technologies used in the project to ensure that nothing is amiss in so far as risk identification is concerned. Where such domain knowledge does not rest within the project team, the Delphi technique should be considered. The Delphi technique involves a group of experts who are queried through several rounds of discussions. Such discussions are conducted anonymously through a facilitator. Anonymity is needed to ensure that senior experts do not unduly influence the views of the less senior experts. Discussions centre around a specific risk scenario and the results are analysed by the facilitator, organized and returned to all the experts for them to provide further views and opinions. The process is repeated until a consensus is reached among all experts. Because the Delphi technique iteratively relies on experts to draw conclusions, the process can be tedious and time-consuming. It is for this reason that in-depth interviews with one or two experts may be a better alternative than the Delphi technique.

Through in-depth interviews, risks are identified by drawing on the knowledge of experts with prior relevant experience. Another comprehensive approach that may be used for risk identification is the SWOT analysis wherein the Strengths, Weaknesses, Opportunities and Threats associated with one aspect of the project are evaluated in details. For example, a new method of construction proposed for the project can be examined for its strengths, weaknesses, opportunities and threats. Through this detailed analysis, the risks associated with the use of this new construction method can be identified in the process. It is always prudent to examine the assumptions made to identify if there are corresponding risks associated with such assumptions.

The reliability of all assumptions made should also be assessed. For example, the subsoil conditions in a large building site should not be assumed to be the same throughout the site based on past geological records of the previous building. Such an assumption will lead to unreliable foundation designs for the new building. Soil investigation and tests should be conducted to determine the soil bearing capacities and soil strata at different depths. Such information serves to identify the risks associated with poor subsoil conditions as well as the corresponding mitigation measures. Similarly, where vendors are concerned, it should not be assumed that all vendors are able to perform equally and therefore a one-size-fits-all approach to risk mitigation can be used for all vendors. Such an assumption is not reliable even if the vendors deal with the same type of materials, have been certified to the same ISO 9001 standards for quality management systems and have been with the project team for a long time.

b. Available tools

The project team can utilize several tools for risk identification. Such tools are typically visual in nature so that project team members can just at one glance have an overall view of what these risks are as well as their causes and effects. One such tool is the fishbone diagram which is also known as the Ishikawa diagram named after its founder in Japan. The fishbone diagram looks like a fishbone, hence its name. It is also referred to as the cause-and-effect diagram. Essentially, the fishbone diagram is like a mind-mapping exercise. It starts with the problem to be resolved. For example, take a problem that is associated with water leakage in the wet areas of a building project. By working backward, the major causes leading to the effect or problem are identified. The contributing causes that lead to each of the major cause are then identified in turn. For example, the major causes leading to problem or effect may be attributed to materials, manpower, scheduling, equipment, availability and workmanship quality.

For the major cause of workmanship quality, the contributing causes could be poor training provided to workers, improper tools, slack supervision and unreliable subcontractors. Having identified all the major causes and their corresponding contributing causes, the project team is then able to obtain an overall view of how all these causes are related to the problem using the fishbone diagram. After studying the fishbone diagram, the appropriate mitigation measures can be taken to resolve the problem or effect. Hence, the fishbone diagram is a useful tool for the project team to identify risks associated with an activity. As a tool that provides a visual representation of the causes, effects and mitigation measures, it helps to provide better understanding as well as greater clarity for the project team to undertake the risk identification process.

Another visual tool that is helpful for the project team to identify risks is the flow chart. The process flow chart shows the relationships between components and activities. By showing the linkages between components and activities, the flow chart is able to show how the overall process works for a specific portion of the work breakdown structure. In highlighting the relationships, the process flow chart is useful for identifying the risks between components and activities. By showing the linkages between components and activities, the flow chart is essentially showing how different parts of a project are related. It is always good to show a total picture of the project using the flow chart for clarity and better understanding of how a delay in one part may cause risks to occur in other parts. A flow chart can be turned into an influence diagram when more concrete details are provided in the chart.

An influence diagram identifies all the components and activities, as well as their variations that may have a spill-over effect on other parts. Such effects can be both direct and indirect. For example, the activities of bricklaying, plastering and painting are typically carried out in sequence. Such activities would have been planned with dates for the bricklayers, plasterers and painters to carry out their work in a sequential manner. If the bricklayers produce defective work, there is a need to rectify the defects which is a direct influence on the bricklaying team. Defect rectification takes time and can lead to a delay in the completion of the bricklaying job. This delay in turn exerts an indirect influence on the start dates of the plasterers and the painters. Hence, visually, the influence diagram is able to show how each activity can influence the others, either positively or negatively.

c. Risk register

Having identified all possible risks in a project, it is necessary to create a tracking system to ensure that such risks are dealt with in a timely manner. This can be facilitated through a risk register which fundamentally forms part of the project plan that documents all information relating to risk management activities. To prevent oversight, the risk register first and foremost contains a list of all the risks that have been identified successfully. These risks can be categorized according to trades such as those relating to the painters, welders, carpenters, roofers, etc. Another category can links to activities such as those relating specifically to scrapping off old paintwork, dismantling formwork, removing asbestos based ceiling boards, etc. The risk register can also be categorized according to elements such as those relating to doors, windows, roof tiles, etc.

Having identified the risks and categorizing them according to trades, activities or elements, the potential responses associated with the risks identified should also be determined. These responses should be included in the risk register together with the risk triggers that provide warning signs that a risk is about to occur or has already occurred. Such triggers should alert stakeholders to immediately take measures to prevent the risks from turning into crises. For example, during the concreting process for floor slabs, there may be risks of formwork collapse due to unforeseen factors. As a risk response, a formwork watcher should be assigned to monitor the underside of the formwork during the concrete casting process. If the formwork watcher observes excessive and unusual bulging of the formwork underside, he should quickly inform the concreters to immediately cease concreting operations. Otherwise, this may lead to collapse of the formwork.

To be able to elicit an appropriate response to the risks identified, it is necessary to first understand the root causes for the risks. The risk register should therefore identify the root causes for the risks alongside the potential responses. For example, causes for formwork collapse can include inadequate formwork design by the structural engineer, poor materials used for formwork erection and poor workmanship by carpenters erecting timber formworks. Finally, the risk register should also appropriately provide an update to the list of risks in the Technical risks, Organizational risks, Project management risks and External environment risks categories. For example, following the introduction of new regulations in the industry, it may now be necessary to reclassify a certain class of risks as "hazardous" as well as identify the persons responsible for managing that risk.

11.4   Qualitative Risk Analysis

a. Subjective nature

It may not always be possible or desirable to quantify risks in the first instance. This is because the perception of risks carries with it a certain degree of subjectivity. Hence, it is more realistic to start with qualitative risk analysis where risks are assessed based on their probabilities of occurrence and their corresponding impact on the project if they do occur. The risks are then scored, ranked and prioritized based on their probabilities and impact. Clearly, the most important risks are those with the highest probabilities of occurrence and highest impact. Where necessary, the conclusions drawn from the qualitative risk analysis may reveal the need for more in-depth quantitative risk analysis or progress directly to planning for an appropriate risk response. Qualitative risk analysis therefore frequently provides a set of preliminary results that surface the need for more in-depth numerical analysis. It is also typically the case that closer to the project completion deadline, qualitative risk analysis is adopted to seek solutions to new risks identified.

This is because the lack of time nearer to project completion precludes the use of quantitative risk analysis which is usually more time consuming to execute. Quantitative risk analysis is time consuming because data needs to be generated and collected for assessment. In a situation where time is a constraint, it is more expedient to make a qualitative risk analysis by consulting the experts. The type, nature and magnitude of a project also have a bearing on qualitative risk analysis. Where the project is a recurring one, reference can be made to historical information for guidance to shorten the learning curve. However, for projects that are new or adopt new technologies, the degree of uncertainty is bound to increase. In such a scenario, qualitative risk analysis may not be suitable. This is because statistical measurements, historical information and even expert knowledge are more reliable than gut feel or a hunch. Nevertheless, a new project or a new technology used today will no longer be new when the project is completed. These can in turn serve to provide historical information on risk management for use in future similar projects.

b. Probability, impact and data precision

Qualitative risk analysis is underpinned in a large way by probability and impact issues. The probability of a risk occurring refers to the likelihood that the risk event may happen. In statistical terms, probability can range between 0 and 1. A probability of 0 means that there is no likelihood of the risk event occurring. At the other end of the spectrum, a probability of 1 means that there is certainly a likelihood of the risk event occurring. Risk impact is the extent of the consequences that the event will cause to the project and its activities. There are two primary approaches to ranking risks.

The first approach is by using the Cardinal scale which is essentially through the use of numerical values. For example, the probability and impact of a risk on the project can be expressed using numerical values from 0.10 representing very low likelihood to 1.00 representing certainty. The second approach is by using the Ordinal scale which is not expressed in numerical values. For example, the risks identified are ranked as "very likely" to "moderately likely" to "very unlikely." Nevertheless, assessment of the risks identified cannot always just rely on qualitative risk analysis. Where the situation permits, it is always more reliable to make observations and take measurements to achieve more reliable results. For example, in a steel cable suspension bridge project, the risk of lightning strikes can have serious effects on the bridge integrity. In this case, it is possible to install automated monitoring devices on the bridge that allows the number of direct lightning hits to be tracked and recorded over a period of time. The data collated helps to facilitate more effective risk response planning to take place.

Following the identification of both the probability and impact of a risk, a probability-impact matrix can be formulated. Probability and impact may be rated subjectively using descriptors such as High, Moderate and Low. If the probability of a risk is low and its impact is also low, then the risk score in this case is low. The evaluation may also be rated numerically. For example, if the probability of a risk occurring is 0.80 and the impact of the risk on the project is 0.10, then the risk score is probability multiply by impact which is 0.08. The risks identified among the activities can then be ranked according to their risk scores which can either be on a subjective basis or numerical basis. Because risk assessment can be carried out based on subjective evaluation or numerical evaluation, the issue of data precision becomes crucial.

This is one of the pitfalls of qualitative risk analysis whose process can be undermined by biasness, subjectivity and preferences. This happens because there are differences in values, beliefs, knowledge, and risk tolerance between individuals. Apart from the subjective nature of qualitative risk analysis, achieving data precision for a start depends on stakeholders' understanding of the project risks. This relates to the level of details and coverage of the risk identified which is also dependent on data availability and access to ready information. Even when such data is available and the information is readily accessible, their quality and reliability should be checked. Otherwise, poor data and information will do more harm than good for risk analysis. Good data however does not come cheap. In addition, better data needs time to collect. The related cost and time issues to collect good data should be factored in for decisions to use qualitative risk analysis.

c. Outcome of qualitative risk analysis

The element of subjectivity can be strong in qualitative risk analysis. The analysis essentially ranks and prioritizes the risks identified based on their probabilities of occurrence and their corresponding impact. By ranking and organizing the risks identified, qualitative risk analysis surfaces risks that require further quantitative risk analysis. Otherwise, the outcome of qualitative risk analysis identifies risks where planning for their risk responses may take place directly, without additional quantitative risk analysis. Fundamentally, the outcome of qualitative risk analysis is to prioritize the risks identified which in most cases should relate to public health and workplace safety as foremost considerations. Qualitative risk analysis may also identify risks that are not critical and only need to be acknowledged and documented for completeness.

This is to ensure that due diligence is taken to account for all risks even though no further action needs to be taken. For example, the risk of trespassing by members of the public into a building site during construction may not be critical if the site is already hoarded and guarded by security officers. The final outcome of qualitative risk analysis may lead to an overall risk ranking for the project. This is over and above the risk rankings of individual activities in the project. The overall project risk score and likelihood of project success can then be compared with other projects. Such comparisons facilitate better understanding of why certain projects are perceived to carry more risks than others, which have implications for corresponding risk mitigation measures such as quantum of insurance premiums. Another outcome of qualitative risk analysis can lead to identification of risk response planning in different categories of risks identified earlier in the project. Such categories can include Technical risks, Organizational risks, Project management risks, and External environment risks. Based on qualitative risk analysis at the activity or category level, critical near-term risks can be identified which demand immediate or near-term risk responses.

These immediate or near-term risk responses are normally associated with activities that are related to occupational health and safety and for which responses cannot be delayed. Immediate or near-term risk responses may also be related to actions needed to comply with new laws relating to public health and workplace safety. Another important outcome of qualitative risk analysis is the identification of risks that are perceived to have low impact on the project and can therefore be placed on a "to watch" list for low priority risks. Otherwise, if the risks are perceived to have high impact on the project, these need to be forwarded for further quantitative analysis. Moving forward, qualitative risk analysis should not only deal with the risks when these occur. A more appropriate response for long-term good would be to examine their root causes with a strategic view to eliminate these risks completely.

11.5 Quantitative Risk Analysis

a. Preparatory work

Quantitative risk analysis is concerned with assessing the probability and impact of identified risks numerically. The end result of the assessment is an overall risk score for the risk identified. Quantitative risk analysis however generally requires more time and costs to complete. A good practice is to first adopt qualitative risk analysis before proceeding to quantitative risk analysis. If a decision can be made after completing qualitative risk analysis, there is no need to proceed further to quantitative risk analysis which is time-consuming.

Quantitative risk analysis starts with the formulation of the research question, data generation, data collection and may require special software for computation of large data sets. The outcome of quantitative risk analysis is to determine the extent of risk exposure and identify risks with the greatest severity to collectively ascertain the likelihood of achieving stipulated project success or objectives. It also considers the corresponding effects of the risks on time, scope and cost targets and if these can be covered using the contingency funds set aside for risk management. Where appropriate, relevant risk management software can be used for computing and analysing the data collected.

b. Sensitivity analysis

One of the key considerations of quantitative risk analysis is to assess the sensitivity of how a change in a variable can affect other variables and to what extent. To do so, it is necessary to evaluate and identify which risks have the greatest severity or impact on the project success. Having identified these risks, management should be alerted for appropriate measures to be taken. Faced with these risks, management needs to determine if more resources should be poured in and by how much to lessen the impact or severity of the risks. If a certain quantum of resources is poured in, then a sensitivity analysis needs to be made to determine how the increase in the use of resources can potentially reduce the severity of the risks and by how much. Sensitivity analysis can be carried out using the concept of Expected Monetary Value.

For example, an identified risk may cost an additional $10,000 to the project and there is 0.30 probability of it occurring. Hence the Expected Monetary Value of the impact of this risk is $3000. Another alternative to managing the same risk may cost the project an additional $20,000 but there is now only 0.20 probability of it occurring. Hence, the Expected Monetary Value of the impact of this risk based on the alternative measure is $4000. There are now two Expected Monetary Values based on two alternative measures; namely $3000 and $4000. All things being equal, the sensitivity analysis would suggest that the former alternative appears to be more attractive with a saving of $1000. In some cases, more alternative measures may be identified. In such cases, decision trees can be used to facilitate systematic assessment of these alternative measures.

c. Decision trees

Where two or more alternatives are available for managing risks, systematic assessment of these alternatives can be facilitated more readily using decision trees. This is a useful method to make the best decision when faced with an array of possible decisions for an identified risk. For example, in the context of procurement, decisions need to be made between make-or-buy alternatives, purchase-or-rent alternatives and outsourcing-or-in-house alternatives. Modelling a decision tree for these alternatives allows the costs and benefits associated with each of the alternatives to be determined.

The decision tree model examines the outcome of each alternative. In decision tree modelling, the best decision is based primarily on the largest value derived from the various possible decisions. The best decision is typically the one with the best cost outcome and highest likelihood of on-time completion. This means that the decision with the largest saving is likely to be the best decision. However, this may not always be the case. In exceptional cases where time is of the essence, the greatest likelihood of on-time completion may be the best decision.

d. Risk response planning

An outcome of quantitative risk analysis provides inputs for planning a risk response for an identified risk. Risk response planning generally focuses on how the probability of identified risks can be reduced so as not to adversely affect the project vision and objectives. For example, the risks of safety lapses have been assessed to be high if timber formwork is used in the project. Risk response planning may suggest closer supervision to ensure that safe practices are adhered to at all times. Alternatively, risk response planning may suggest replacing timber formwork with metal formwork which is more stable. Nevertheless, not all risks are negative risks. When faced with positive risks, the opportunity should also be taken to enhance the likelihood of positive risks for project success. For example, currency hedging when carried out properly by finance professionals to guard against foreign exchange fluctuations can be a positive risk.

Another example could be when two cost consultancy firms who are traditionally rivals decide to enter into a strategic alliance to pool their resources to take on a mega project which none of them would be able to take on individually. While there are risks relating to revelation of trade secrets and sharing intellectual property rights, the benefits of them working as team may far outweigh the perceived risks. The responsibilities for making this happen in order to capitalize on the positive risks should be assigned to the right people and teams who work closely with the risk event. Risk response planning should however be realistic and practical. Over-kill and gold-plating should be avoided. For example, the risk identified corresponds to a problem that would cost the project $5000 if it materializes. In this context and provided safety is not compromised, it is not realistic nor practical to spend an additional $500,000 trying to solve this problem! At the end of the day, risk response planning aims to lower the overall project risks to a level and at a cost that is acceptable.

11.6 Ways to Manage Risks

a. Avoiding negative risks

In situations where negative risks have been identified, the best way to manage the risks is simply to avoid them. Avoiding negative risks can be achieved in several ways. Firstly, the project plan can be changed or modified to avoid the risk. For example, if a particular way of working carries negative risks, then that way should be changed. Likewise if the use of a particular technology has negative risk connotations, that technology should be replaced. Secondly, negative risks may occur because of miscommunication or misinterpretation. Hence, any chances for miscommunication and misinterpretation to happen should be eliminated through clarification or to simply ask for more information.

Thirdly, negative risks can be present if there is a manpower crunch in a project leading to overwork and safety lapses. The solution to avoiding this is to employ more workers to even out the workload to prevent fatigue. Likewise, if the project team members lack the necessary experience or expertise, this may heighten negative risks. Hence, efforts should be made to hire people with the relevant experience to avoid the negative risks. Lastly, negative risks can also be avoided by simply taking the well-trodden path where the learning curve has already set in rather than to heighten such risks by using a new method of working which the project team members are not familiar with.

b. Transferring negative risks

If the negative risks cannot be avoided, then consideration should be given to transferring such risks to a third party. This also transfers the ownership of the risks to the third party. The transfer of such risks to a third party to own and manage the risks would require the payment of a premium by the party transferring the risks. In such a transaction, the risks are still there except that these now become someone else's problems. The party taking over the risks is typically the insurance company. There are several ways in which negative risks can be transferred. The first and most common way to transfer risks is through insurance. In this way, the party transferring the risks pays a premium to the insurance company who then underwrites the risks. Risks can also be transferred through performance bonds. In this approach, the client may stipulate that the contactor obtains a performance bond from the bank in consideration of paying a fee. In the event that the contractor fails to complete the project, the performance bond will be called upon wherein the bank compensates the client.

Warranty is another way where risks can be transferred. For example, in return for a payment, an extended warranty can be purchased to safeguard the good working conditions of an electrical fixture in a building for a stipulated period of time. If that fixture breaks down, the party that issues the warranty has to undertake repair works to fix the problems. The warranty can be further extended upon expiry of the warranty period. Another approach to transferring risks is to obtain a guarantee. However, unlike a warranty, a guarantee is normally for a fixed period of time and is not renewable. The use of fixed price contracts is also able to transfer risks from the client to the builder. This is where the builder receives a fixed lump sum to complete the project regardless of how the project will turn out to be. If there is overspending, the builder is not able to request for more money from the client. By locking into a fixed price, the client effectively transfers the risks to the builder.

c. Mitigating negative risks

The mitigation of negative risks means taking steps to prevent the risks from happening. This requires efforts to reduce the likelihood and severity of the risks identified. Such efforts should commence even before the risks start to surface. The rationale for mitigating risks is that prevention is always better than cure. This is because preventive costs are generally less than corrective costs. Consequently, the time and costs required to minimize or eliminate the risks should be more cost effective than rectifying the damage caused if the risks materialized. In addition, such damage may hurt the reputation of the organization. Mitigation measures can be taken in various ways. A simple way is to enhance the activities to reduce the likelihood or severity of the risk. For example, this can take the form of increasing site patrols to reduce trespassing or enhancing the frequency of safety checks to prevent accidents from happening. Another way to mitigate risks is to make the work processes as simple as possible. When workers clearly understand what they need to do because the processes are simple, this effectively helps to reduce the risks of making mistakes and safety lapses.

Other mitigation measures can include the completion of more tests, conduct simulations or to develop and test-bed prototypes before full system implementation. Trial runs can also be carried out as a mitigation measure before the actual event takes place. For example, testing of electrical installations in a new building project should be carried out before inspection by the authorities takes place for the purpose of commissioning the installation. A purposeful control of supply or limited releases is also another way to mitigate negative risks. Limited releases allow products to be tested to be defect-free before a full launch is made. In the event that there are defects, these would only be confined to a small number of such products, thus dampening the negative risks. For example, in a new 600-unit condominium project, the developer may choose to launch only 50 units for sale in the first phase. The sale of a small number of units tends to be easier and allows the developer to test the market. This also mitigates the risk of negative publicity if the project fails to attract buyers if more units go on sale all at the same time.

d. Managing positive risk opportunities

Not all risks identified in a project are negative risks. Where positive risks are present, these should be exploited by the project team members to their advantage. For example, quality enhancements to a new condominium project may risk having substantial costs increase. However, this can be offset by increased sales and better customer satisfaction which should correspondingly be exploited. Positive risks should also be shared where possible among partners. New partners may need to be established if the positive risks are too large for one party to shoulder. For example, if a company is unable to take on a very large project on its own, it should consider partnering another company to bid for the job. Large projects offer much opportunity not to be missed and such opportunities should therefore be shared if necessary. Furthermore, the positive risks identified should also be enhanced where possible to realize more benefits. In cases where the positive risks need to be activated, their respective triggers have to be identified to ensure that the positive risks do indeed happen so that the corresponding rewards can be reaped. The project team must make it happens. For example, instead of just building for others, the contractor may diversify upstream into property development which is more lucrative.

e. Accepting the risks

Not all risks identified need to be transferred, avoided or mitigated. There are risks which can simply be accepted because transfer, avoidance or mitigation is not viable. Risks that can be accepted are typically risks that have low probability and severity. Hence, such small and insignificant risks can be accepted without the need for any formal risk response. Nevertheless, no matter how small the risk is perceived to be, monitoring is still necessary to ensure that it does not degenerate into a larger risk if left alone. For example, cast in situ concrete placing may be planned for a specific date and time. The ready-mixed concrete supplier has been notified to make deliveries at the specific date and time. At the time of planning, it is not known if the weather will be good for concrete casting to take place. There is a remote possibility that it may rain on that day which can hinder concrete casting. While reference can be made to weather forecasts, the likelihood of bad weather is always there. This risk can be accepted but with provisions made for possible postponement of concrete casting if it rains heavily. This is known as passive acceptance where no action is required and the project team deals with the risk only when it happens.

On the other hand, active acceptance means a contingency plan is put in place to deal with the risk identified if and when it happens. Continuing from the concrete casting example, a contingency plan should already be in place to continue operations after the rain stops. The contingency plan can include having the concreting team on stand-by to wait out the rain and advanced notification to the ready-mixed concrete supplier to postpone the deliveries of concrete to site. Funds should however be set aside to operationalize the contingency plan. This is known as the contingency allowance. The amount to be set aside for this purpose depends on the probability and severity of the risk vis-à-vis the expected monetary value of that risk event. The contingency allowance covers more than one risk event. Some risks may however yield negative impact while other risks may yield positive impact. For example, Risk A has a probability of 20% and a negative cost impact of minus $4000. Its expected monetary value is therefore minus $800. Risk B has a probability of 10% and a positive benefit of $2000. Its expected monetary value is therefore $200. Hence, the contingency allowance to be set aside is $600.

f. Residual and secondary risks

The world that we live in is not a perfect place. Hence, despite all that we may have done for risk planning, some residual risks may continue to linger on. In addition, not all such residual risks can be eliminated through mitigation, avoidance or transfer. On a positive note, such residual risks are typically minor in nature and can be accepted. For example, when a new building project is completed, cleaning of the premises is necessary before handing over to the client. The cleaning tasks may be delayed with the result that hand over is late. Nevertheless, given that cleaning is the last task to be carried out at the tail end of the project, it may be considered to be a residual risk even if there is likelihood of late completion. Another class of risks is secondary risks which may be triggered by the risk responses undertaken earlier. For example, risk transfer to an insurance company carries with it a secondary risk of the insurance company going into liquidation. Organizations that subscribe to cloud computing for storing their data also face the secondary risk of cyber-attack if the cloud computing facility is not properly secured by the service provider.

11.7   Monitoring and Controlling Risks

a. What this process entails

When the project progresses, risk monitoring is important to ensure that signs of the identified risks occurring are noted early. Actions are then taken to control the risks using the risk response plans agreed earlier. When the risk response plans kick in, continuous monitoring also takes place to detect if new risks may surface. The outcomes from the risk response plans should also be documented for future reference for both success and failure cases. Records should also be kept of the metrics or symptoms used to show if the risks are happening, tapering off or coming to an end. One example in this context can be related to the monitoring and controlling of currency exchange risks associated with the purchase of materials from an overseas vendor. Using geotechnical sensors to monitor ground movements during deep excavation work is an example of how technology can be harnessed to record if ground movement is happening, slowing down or cease.

The objectives of risk monitoring and control is to ensure that when identified risks occur, timely responses are taken and that the outcome is effective as planned. During risk monitoring and control, the earlier assumptions made about the risks are also tested to determine if these are valid. In addition, the risk monitoring and control process tracks changes in risk exposure, risk triggers and the appearance of new risks. The process essentially monitors and controls existing risks and spots new risks.

b. Completion of risk monitoring and control

Several approaches can be used to support risk monitoring and control in progress until its completion. Firstly, auditing risk responses is a useful approach to evaluate the planned risk responses. The audits provide opinions of whether the planned risk responses are justified, effective and reasonable. Secondly, every project team meeting should include periodic risk reviews in the meeting agenda. Such periodic reviews allow the project team to monitor and control risks on a regular basis. The reviews also help to facilitate continuous improvements to the risk response plans if appropriate.

Thirdly, where data is available, earned value analysis can also be conducted to measure project performance focusing on monetary outcomes based on the risk response plans agreed earlier. Fourthly, technical performance can also be measured if the risk response plan is related to a specific technology used. The measurement should be able to highlight the technical competence of the project team relating to the use of this technology. The results serve to provide a closure to the risk monitoring and control exercise for this technical area of work. Finally, completing risk monitoring and control may not necessarily lead to completion per se. Instead the completion outcome may give rise to additional risk planning to watch out for new risks.

11.8 Revision Questions

1. Is there a difference between recklessness and informed risk management?
2. What is the difference between risk and crisis?
3. What is risk management planning about?
4. How is the stakeholders' tolerance for risks assessed?
5. What areas of responsibilities does a risk management plan cover?
6. How does a risk breakdown structure work?
7. How are risks identified?
8. Why is it useful to use the Ishikawa Diagram for risk management?
9. Why is it useful to use flowcharts for risk management?
10. Why is it useful to use influence diagrams for risk management?
11. Why is it necessary to create a risk register?
12. What does qualitative risk analysis entail?
13. How are risks ranked and prioritized in qualitative risk analysis?
14. What is risk probability?
15. What is risk impact?
16. What are cardinal scales used in qualitative risk analysis?
17. What are ordinal scales used in qualitative risk analysis?
18. Why is the probability-impact matrix useful for project risk management?
19. What are some of the concerns relating to data precision in qualitative risk analysis?
20. What are some of the end results of qualitative risk analysis?
21. How is quantitative risk analysis to be prepared?
22. Why is sensitivity analysis important for analysing the Expected Monetary Value?

23. What is the usefulness of decision trees?
24. Why is planning for risk response an essential part of project risk management?
25. What are some of the methods used to avoid risks?
26. How are negative risks transferred?
27. How are negative risks mitigated?
28. What are positive risks and why should these be exploited?
29. When should risks be accepted?
30. Why do projects need to provide for contingency allowance?
31. What are residual risks?
32. What are secondary risks?
33. Why is risk monitoring and controlling required in projects?
34. What are some of the tools available for risk monitoring and controlling?
35. Why is quantitative risk analysis more costly than qualitative risk analysis?
36. How is project risk management different from enterprise risk management?

Chapter 12
Project Procurement Management

12.1 Nature of Procurement

a. What, why, how and when to procure

Projects require a variety of resources for their completion. These resources however do not simply appear out of nowhere. They have to be procured or purchased. The project manager needs to identify and plan for what resources are necessary for the project. Depending on the nature of the project, resources can include machineries, equipment, tools, materials, tradesmen, supervisors, consultancy services, employee training as well as a host of other goods and services. Procurement management is the process where-in the project manager plans for and purchases the goods and services required to achieve the needs identified in the project scope. He also seeks out potential vendors who can supply the goods or services needed at a price that is fair and reasonable as well as able to meet the time, cost and quality criteria of the project.

Furthermore, procurement management needs to consider if the resources required should be purchased from external vendors or made in-house. Planning is important for procurement. The first question to ask is what does the project needs. Sources where goods and services can be procured have to be identified. For a specific good or service, there may also be more than one source where the purchase can take place. Criteria have to be set and assessed to decide who is the best vendor for the good or service. After a vendor has been identified and selected, an agreement needs to be signed with the vendor to ensure that all contractual obligations are set out clearly and delivered in full. The contract needs to be administered properly especially for procurement that spans over a long period of time. After the good and service as procured have been delivered and accepted, the contract with the vendor will need to be closed.

© Springer Nature Singapore Pte Ltd. 2018
L.S. Pheng, *Project Management for the Built Environment*,
Management in the Built Environment,
https://doi.org/10.1007/978-981-10-6992-5_12

b. Procurement planning

The project manager identifies the resources needed to deliver the project through the scope statement. Different types of resources are identified including their quantities and time-line for their deliveries to the building site. Some resources such as concrete can be purchased from more than one supplier to avoid out-of-stock situations. It is always useful to rely on existing templates of similar completed projects to check on the types of resources typically required for such projects. Such templates provide the project manager with a useful checklist for procurement planning. For each part of the project, the project manager needs to identify whether procurement is necessary and if so for what resources. The quantities of such resources have to be determined and dates of their orders, reminders and deliveries to the building site included in the procurement plan. From the perspective of the entire project, procurement planning involves all components and activities identified in the work breakdown structure.

In addition, the project manager needs to consider current as well as future marketplace conditions to assess the risks associated with procurement. Current marketplace conditions may suggest that the resources required are better off by making these in-house because the organization possesses the expertise and manpower for this purpose. If the project stretches over a few years, future marketplace conditions may indicate inflationary effects, foreign exchange fluctuations and obsolescence affecting procurement planning. The materials currently identified for a project may no longer be available in the future when these are needed or may be superseded by other more advanced alternatives. Having identified the resources required, the mode in which procurement takes place has to be considered. For example, in public sector projects using taxpayer monies, it is often a requirement to request for at least three quotations from different vendors. All things being equal, the vendor quoting the lowest price is normally chosen for the job. The procurement process should also follow established organizational policies and procedures to ensure transparency and arms-length transactions that are void of potential conflicts of interests.

c. The marketplace

Procurement planning frequently focuses on "make" or "buy" decisions. The decision to buy therefore needs to consider the marketplace within which the transaction takes place. The marketplace can influence procurement planning as well as the procedures adopted to select vendors. Typically this depends on the nature of the good or service to be procured and if there are only one or many sellers in the marketplace where the purchase can be made from. In the first instance, the nature of the good and service may be such that there is only a sole source in the marketplace with only one seller who is qualified to do so.

This may happen because the one qualified seller possesses a unique technology or expertise to produce the good or service required. The required good may also be patented by the sole qualified seller. Such unique situations should be noted in the procurement plans especially when the normal procurement procedures stipulating the need for at least three open tender are waived. There can also be other situations where the client or the project team prefers to work with a specific vendor. This arises because the preferred vendor has successfully worked with the client or project team before and has built up good relationships with excellent track records. In addition, the preference may come about because it is easier for the project designer to incorporate what the specific vendor has to offer into the project. The preference may be indicated by the client or the project designer or recommended by the project team. There may also be instances where a mega building project involves a client with related subsidiaries who supply materials such as ready-mixed concrete.

The client may consequently stipulate that concrete for the mega building project be sourced only from that one subsidiary producing ready-mixed concrete. Contractually, the preferred vendor is often referred to as the nominated supplier or nominated subcontractor. Decisions to buy can also be influenced by the number of vendors in the marketplace producing a specific good. When there are a large number of vendors, prices are likely to be more competitive as free market forces kick in. On the other hand, there may also be very few vendors in an oligopolistic market such as those in the telecommunications sector or firms producing precast concrete staircases. In such a setting, the price set by a vendor is likely to directly affect the pricing of other vendors and hence the overall marketplace conditions.

12.2 Identifying Resources for Procurement

a. Scope statement

Identifying the resources needed for the project starts with the scope statement. It should be reiterated that the scope statement is not a mere statement of a few paragraphs. Depending on the scale and complexity of the project, the scope statement can be a set of hefty documents setting out in details what is needed and therefore need to be purchased for the project. The work breakdown structure is derived from the scope statement from where the various components and activities are identified to achieve the end results of the project. These documents provide inputs to the project management plan which shows the times when various purchases are to be made. Timing in the procurement context is more than just the time when deliveries are made to the project site. Timings for procurement include the lead time for placing the order, confirming the order, sending reminders and notification for delivery on the actual day. For large complex building projects, procurement management works hand-in-hand with time management to determine the various timings for all the components and activities identified in the work breakdown structure.

b. When to start procurement planning

In view of the large number of purchases that a building project is expected to make, it is always desirable to start planning for procurement as early as possible in the early stages of the project planning process. Purchases are made for resources that are directly needed for the project. Examples include claddings and ready-mixed concrete. However, from the method statements prepared by the builder, it is also known that a crane is required for hoisting and installing the claddings. Similarly, a concrete pump is also required to transport the ready-mixed concrete to upper floors for casting. Hence, in this context, procurement planning is not confined only to the claddings and ready-mixed concrete. Procurement planning should also account for the crane and concrete pump.

Procurement planning should be done as early as possible because vendors need time to prepare their quotations as well as to ready themselves to supply the goods and services contracted for. Lead time for placing orders should also be factored into procurement planning. While procurement planning should start as early as possible, it is noteworthy that planning for procurement is an on-going process that is required throughout the entire project duration. This occurs because clients or designers may order changes midway through the project due to budget constraints or availability of better building materials. Changes, known typically as variation orders in building projects, tend to be more frequent in the early stages of a project where the implications arising from a change are not that adverse. However, as the building project moves closer towards completion, changes can be costly and undesirable because of possible mismatch.

Not all changes are caused by clients and designers. There may be instances where the change comes about because of the introduction of new regulations such as the requirement for a new health and safety supervisor to be positioned full-time on the building site. These scenarios mean that while procurement planning should start as early as possible, this can continue throughout the project. In both cases, procurement planning should also start with determining if a "make" or "buy" decision should be made. Even in the "buy" context, there can be further options. For example, it is not possible or desirable for a construction firm to make its own mobile crane. The firm can however decide to either buy a new crane or a second-hand crane, rent or lease the crane from a plant hiring company. Once a "buy" or "make" decision has been made, it is also a good practice to document the reasons for future reference. For example, in the case of a confined building site in the central business district, most procurement decisions would hinge heavily on buying instead of making because of the severe lack of working space on site.

Similarly, if a large construction firm owns a subsidiary that produces ready-mixed concrete, it is to the firm's benefit to buy concrete from the subsidiary than to batch the concrete on site. In exceptional cases involving unique technologies or methods of construction for which the project manager does not have the relevant experience, he is more likely than not to rely on technical experts to provide him with advice on procurement decisions. This may relate to for example extremely poor geological conditions and the supports required for the excavation for basement construction. The reliance on expert inputs and judgement should also be documented as part of procurement planning for future reference.

c. Do we make or buy?

There are many factors to consider when deciding if resources needed for a project should be made in-house or purchased from external vendors. The first factor to consider is cost. Cost is certainly an important consideration wherein the cheaper option is often selected, all things being equal. For example, a builder is likely to buy precast concrete components from a precast concrete supplier. There is little or no choice for the builder here as it does not make sense for him to incur heavy capital investments to set up a prefabrication yard for this purpose. There is no economy of scale for him and the manufacturing of precast concrete components is also not his core business.

The second factor to consider is whether there are existing in-house skills and expertise to make the resource required for the project. For example, the builder may wish to implement a project management information system in his organization. However, he does not have any information technology staff in his organization that possesses the expertise to develop the system. Hence, the best option is for the builder to outsource the development of the system to an external information technology vendor. If such a system already exists and can be bought readily off the shelf, this should lead to a "buy" decision. The third factor relates to the extent of control which the builder wishes to exercise in-house. This can be related to intellectual property rights where confidentiality is better preserved if the innovation is made in-house. The fourth factor can be linked to staff development where "make" decisions allow employees to learn new trades, skills and expertise by doing things in-house. Employee involvement and participation can help lead to better motivation and staff morale.

On the other hand, there are also valid reasons to purchase from vendors. "Buy" decisions make more sense if it is less expensive to do so and there is no existing in-house staff with the skills or expertise to render the "make" decision feasible. Buying from vendors is thus recommended if it is more efficient to do so especially if the order involves small quantities and there are ready vendors available. This allows members of the project team to focus on more important tasks instead of being straddled with small volume work. This also effectively helps to transfer the risks from the project team to the external vendor. Finally, apart from initial costs, consideration should also be given to servicing costs during the lifespan of the good or service procured. A specific product may be cheaper to purchase from a vendor initially but is expected to incur more expensive servicing costs compared to a comparable product that is made in-house. Hence, a balance needs to be struck between the "buy" or "make" option.

12.3 Contractual Issues

a. What are procurement contracts?

Having agreed to an agreement for the supply of goods or services, a contract needs to be established between the buyer and the seller. A contract can either be made orally or formally in writing. However, a contract should preferably be presented in a formal document in writing. The written format helps to avoid ambiguities and make for clear understanding of the terms and conditions of the agreement. A written contract can be drafted as a new document. Alternatively, parties can agree to use a standard form of contract for this purpose. Examples of such standard forms of contract used in the construction industry are the Public Sector Standard Conditions of Contract and the Singapore Institute of Architects Standard Conditions of Contract for Building Works.

Where new contract documents are to be drafted or existing standard forms of contract modified for use in a project, it is best to seek legal counsel to ensure that the agreement signed by parties is enforceable in the courts of law. The contract should set out the terms and conditions clearly for product acceptance in the project. The procedures leading to acceptance can include inspections, reviews, audits and walk-throughs. Upon signing the agreement, future changes to the contract must be agreed by the parties concerned and be approved formally. Such changes can include variation orders and change control as the project progresses. Such changes must also be properly documented and controlled for future reference.

b. What contracts entail

All requirements set out in the contract must be fulfilled. For example, if a new building project is unable to successfully obtain a temporary occupation permit from the authorities upon completion, the contract is not considered to have been fulfilled. The shortcomings identified have to be rectified in a timely manner for the permit to be obtained before the contract is fulfilled. Hence, contracts must set out clearly what are breaches, intellectual property rights and issues relating to copyrights. In additions, contracts should also provide for unforeseeable events that cannot be reasonably anticipated by parties such as those caused by exceptionally inclement weather and natural disasters. It should also be noted that in a typical building project, the contractual framework is not confined only to the builder and the client.

There are also contracts between the builder and vendors for goods and services. For example, there are insurance contracts that the builder can use to mitigate risks by transferring such risks to the insurer by paying a premium. There are certain conditions that a contract must meet for it to be valid. Firstly, there must be an offer and an acceptance. Secondly, a valid contract must provide for consideration which is normally in the form of payment for the goods delivered or premium for the insurance. All contracts must be for a legal purpose for them to be valid. For example, a "contract" for providing financial services for the purpose of money laundering is not a valid contract since it is not for a legal purpose. Contracts must also be executed by parties with the capacity and authority to be a signatory to the contract. Such persons must also be of sound mind.

12.4   Types of Contracts

a. Fixed price contracts

In procurement management, parties can choose from a range of contractual terms and conditions depending on which party is willing to take on more risks for a price. Fixed price or lump sum contracts are commonly used in the construction industry. In fixed price contracts, a lump sum is paid by the buyer to the seller. It defines the total price for the goods or services that the seller provides. Fixed price contracts transfer the risks to the seller. This is because the seller undertakes the risk of cost overruns. For example, if a builder agrees to construct a house for the client for a lump sum using a fixed price contract, then the builder only receives that lump sum even if he overspends because of inflation or unexpected increase in material costs. In a fixed price contract, the builder is not entitled to ask for additional payments even if his expenses exceed the lump sum agreed earlier. It is therefore important for the seller-builder to ensure that his cost estimates are realistic, accurate and free from mistakes. He should also factor the cost of risk mitigation into his price. The buyer-client therefore bears less risks in a fixed price contract.

b. Cost reimbursable contracts

Unlike the fixed price contract, cost reimbursable contracts provide for adjustments in the payments to the seller. There are three different ways where such adjustments can be made. The three types of cost reimbursable contracts are the Cost plus fixed fee contract, the Cost plus percentage of costs contract and the Cost plus incentive fee contract. Cost reimbursable contracts are generally used when details of the building project are not known when the project commences. Cost reimbursable contracts are also used when the work is of an urgent nature where the builder is instructed to start work as soon as possible. When the urgent work is completed, the builder is reimbursed for his costs incurred plus an adjustment depending on the agreed formula. In general, when payment is made to the builder, the payment includes a profit component. The profit component is the difference between the actual costs of the goods or services provided and the sale price. It is therefore important for parties to understand what the various costs are from the seller's perspective.

   Actual costs comprise both direct costs and indirect costs. Direct costs include among other things costs incurred for salaries, plant, equipment, materials, etc. Examples of indirect costs include office rentals, utilities, shared telecommunications systems, overheads, etc. In cost reimbursable contracts, the buyer takes on the risk of cost overruns. However, the extent of this risk depends on the type of cost reimbursable contracts used. In the cost plus fixed fees contract, the seller is reimbursed the costs incurred plus a fixed fee that is agreed in advance. In such a situation, the risk to the buyer is lesser compared to other cost reimbursable contracts. In addition, in the cost plus fixed fees contract, there is no incentive for the seller to prolong the project. This is because the fees quantum which the seller receives is fixed regardless of the project duration. Hence, it is to the seller's interest to complete the work as quickly as possible, collect the fees and then move on to other projects.

The cost reimbursable contract that has the highest risk for the buyer is the cost plus percentage of costs contract. In this case, there is no incentive for the seller to rein in costs. This is because the more the seller spends the more he receives in fees which vary depending on the amount spent. Measures should therefore be put in place by the buyer to ensure that the amount spent is not excessive or unreasonable. Such a contract should include a "total not to exceed" clause to safeguard the buyer's interests. The seller therefore assumes less risk in the cost plus percentage of costs contract.

Another form of cost reimbursable contracts does however provide for savings to be shared between the buyer and the seller. This is the cost plus incentive fees contract where savings for the project are shared between the buyer and the seller according to a formula agreed earlier. The sharing formula can be on a 50:50 basis or 70:30 basis as agreed between the buyer and the seller. The cost plus incentive fee contract therefore incentivizes the seller to look for cost saving opportunities. These can come about through better methods of working or introduction of more productive operations without deviating from the project's vision. The more cost savings the seller achieves, the more monetary incentives he receives. This contractual arrangement also benefits the buyer because the cost savings translate to savings for the project.

c. Schedule of rates

Procurement can also be made on the basis of a schedule of rates. This is essentially a schedule that sets out the rates for different items of work needed for a project. For example, the schedule can include rates of $800 for a solid timber door, $56 per m^2 for ceramic floor tiles, and $500 per hour for the services rendered by an experienced professional engineer. All items of work that are necessary to deliver the project should be included in the schedule of rates. The procurement mode using a schedule of rates can also be referred to as Unit price contracts or Time and Materials contracts. This is because the schedule provides for unit price or rate for items of work relating to time expended for labour or materials used. A contract based on schedule of rates is useful for small projects involving smaller amount of work.

A typical example where such a procurement mode is used is the Term contract in local or town councils. Term contracts provide for small renovation and repair works in large estates managed by town councils. Such small renovation and repair works do not happen frequently and are usually one-off in nature. For this reason, the rates submitted by the contractor tend to be higher because there is no economy of scale in such works. However, in a scenario where the volume of such works increases unexpectedly for some legitimate reasons, using such "inflated" rates benefit the seller. From the buyer's point of view, the costs incurred for the contract can spin dangerously out of control as more work is given to the seller. The buyer should therefore include a "total not to exceed" clause in the Time and materials contract as a mitigation measure.

12.5 Procurement Management Plan

a. Items to include

The items to be included in the procurement management plan provide comprehensive guidelines and procedures on how procurement is to be carried out. Having identified the resources needed, the procurement management plan provides guidelines on how appropriate vendors are to be identified and selected. Such guidelines can include the need to prequalify potential vendors for their track records before inviting them to submit bids. In addition, the guidelines may stipulate that potential vendors must be registered with the competent building authorities for public sector construction projects. The types of standard conditions of building contract may also be prescribed for use in the procurement management plan. Examples of such standard conditions of building contract can include the Public Sector Standard Conditions of Contract and the Singapore Institute of Architects Standard Conditions of Contract.

Where such standard conditions of contract also provide for standard procurement forms, these should also be used. For each of the major activity in the project, guidelines should also be provided to guide the process of independent estimating. This would require the buyer to work out his own estimates for comparison with the estimates submitted by potential vendors. Through the comparison, the buyer is able to evaluate if the estimates submitted by potential vendors are fair and reasonable. In addition, the requirement to obtain at least three quotations from different vendors is also commonly prescribed as part of the procurement process. For larger organizations involved with large complex projects, it is not uncommon for these organizations to have central procurement offices to handle the purchase of resources in bulk to benefit from better discounts given by vendors. The reporting and working relationships between the central procurement office and the project team should be spelled out clearly in the procurement management plan.

The plan should also prescribe the manner in which vendors are to make their deliveries to the site to avoid congestion. This is especially critical for confined sites in the busy city centre where traffic is heavy and there is little or no space for delivery vehicles to manoeuvre. In such a situation, the procurement management plan should adopt the just-in-time principle where deliveries are made just-in-time when the resources are needed. Unnecessary waiting time and site storage can therefore be avoided. Project integration and communications management should also feature in procurement to ensure proper co-ordination between the vendors and the project team. This pertains to matters such as timings of deliveries to site and the availability of hoisting facilities on site for installation.

b. Description of work

As part of the contract, a detailed description of work should be provided by the vendor to fully describe the work to be completed or the goods to be supplied. This description of work should form part of the contract between the buyer and the vendor. It should set out clearly if the work involves the supply of the materials only or both the supply and installation of the materials in the project. In preparing the formal contract between the buyer and the vendor, reference should be made to the procurement management plan to indicate the timings of the deliveries and in what quantities. Apart from incorporation of the description of work, other planning inputs to facilitate procurement from the vendor should include the relevant schedules of deliveries, the corresponding quantities as well as the cost estimates.

The constraints if any associated with the deliveries should also be highlighted. For example, deliveries could be contingent on the availability of a tower crane for unloading the materials or subject to traffic control in a busy thoroughfare. Similarly, if assumptions have been made relating to the deliveries, these should also be highlighted for follow-up actions to be taken. For example, the delivery of ready-mixed concrete to the building site assumes that the concreting team is available for casting concrete on that same day. The project team should indeed follow up to ensure that timely actions are taken to turn the assumption into reality.

c. Procurement documents

When making preliminary enquiries to request for a proposal or a price, relevant information should be provided by the buyer to the vendor to facilitate the transaction. As part of the information to be provided, the buyer should include the description of work, drawings, models and specifications to the vendor. If the project involves confidential issues, a non-disclosure agreement should also be signed by the vendor prior to releasing such sensitive information to him. Upon receiving such information, there should also be provisions for the vendor to suggest other viable alternatives with supporting justifications such as further time and cost savings. Organizing such information should be the responsibility of the procurement office either at the project level or centralized at the head office. The procurement office identifies the standard conditions of contract to be used as well as prepare the tender documents. Where necessary, the procurement office should also source for expert judgement for areas of work which the project team is unfamiliar with or to evaluate the best type of and source for the goods to be procured. In this context, consultation with the appropriate experts is required to evaluate and select the new and specialized technology used for the first time.

The commonly used terms in procurement include bid or tender, quotation and proposal. An invitation for bid or tender requests the vendor to provide a price for the good or service to be procured. For example, a contractor is invited to submit a bid or tender for the construction of a two-storey dwelling house. A request for quote requests the vendor to similarly provide a price for a component to be procured. For example, a door supplier may be invited to submit a quote for the supply of a fixed quantity of a particular door type. In the construction industry, a request for quote is generally targeted at the subcontractor or supplier. A request for proposal is generally of a larger scale wherein the vendor is requested to submit a proposal to complete the work procured. For example, a contractor is invited to submit a proposal for the design, construction and operation of a hotel for a country club. Hence, a request for proposal is generally for a total package of work. In all three categories of bid, quotation and proposal, the procurement documents would typically include a caveat that the client or buyer reserves the rights not to accept the lowest or any bids without providing any reasons.

12.6 Evaluating the Vendors

a. What evaluation criteria to use?

When more than one vendor submits a bid or a quotation, some form of evaluation criteria should be used to assess them prior to selection. More often than not, a primary selection criterion is based on the lowest price alone. Apart from price, additional criteria should however be used to evaluate the vendors using a scoring system. Such additional criteria can include relevant experience, references provided by past clients, certifications for quality, safety and environment standards, etc. The scoring system is qualitative and can be subjective in nature. For example, in the case of relevant experience, a score of 1 can be assigned to a vendor without the necessary experience and a score of 10 can be assigned to a vendor with many years of relevant experience needed for the current project.

Depending on how exhaustive one would like the scoring system to be in evaluating the vendors, other considerations can include their reputation, availability of resources, financial capability, suitability, price, cooperativeness, availability and management competence. There may also be existing government policies for public sector building projects that stipulate the mandatory use of the price-quality method to evaluate vendors for contractual values above a certain amount, say $2 million. In using multi-factors to evaluate vendors, it may not therefore always be the case that the vendor who submits the lowest price will win the bid.

b. Update description of work

When vendors build up their bids, quotes or proposals for projects, situations can arise where the vendors offer alternatives to the description of work which they received from the project team. Such alternatives can come about because the vendor possesses the know-how which the project team was not previously aware. The vendor may also undertake a re-engineering exercise to produce an alternative to the existing description of work. If the alternative proposed by a vendor is better than the existing description of work, then it makes more sense for the project team to accept the alternative. In this context, a new description of work is prepared for the alternative proposal. The bid, quotation or proposal submitted by the vendor is then evaluated with this new description of work in mind. Once accepted, changes to the description of work should be updated, documented and recorded with the reasons for the change explained clearly.

c. Preparatory work for contracting

When the description of work is ready, procurement documents should be prepared for inviting vendors to submit bids or requests for proposals. The platform for shortlisting potential vendors for the project team to send out invitations should also be identified. The project team may already possess lists of qualified vendors with the relevant past experience. Their contact details can be obtained from the lists. The qualified vendors are then invited to participate in the bids, quotations or proposals. Alternatively, if the project team does not possess any lists of qualified vendors, such lists may be obtained from the relevant building authorities who for example maintain a central registry of public sector contractors or vendors. The central registry will normally classify vendors into various categories according to their financial grades from the largest firms to the smallest firms.

Such a classification allows the project team to prequalify appropriate vendors for the job. Prequalification means that a small firm should not be invited to bid for a large complex project which the small firm clearly cannot handle. The project team may also develop a prequalification list exclusively for the current project. For public sector projects, there is normally also a requirement to advertise the pre-qualification exercise, invitation to bid and request for proposal in the local newspapers, popular trade magazines or on-line portals maintained by the government. The purpose for advertising is to ensure public accountability in using taxpayer monies. After the potential vendors have been identified, meetings should be held with them to ensure that they have a clear understanding of what goods or services the project team is procuring for the client.

Such meetings also provide the opportunity for the vendors to ask questions and seek clarification for them to ensure that their submissions are appropriate and sufficient to meet the buyer's requirements. It is also a good practice for the project team to meet all the potential vendors at the site on a specific date and time for a site walk-about. The walk-about provides potential vendors an idea of the magnitude of the project. For example, term contractors bidding for local or town council jobs for estate management and maintenance should appreciate from the site walk-about what is expected of them for the project. The site walk-about also provides opportunities for potential vendors to ask questions and seek additional information.

12.7   Selecting the Vendors

a. Preparing to select vendors

Following the receipt of the description of work and invitation to bid, the vendors submit their proposals to the project team for evaluation. The proposals include the bid price as well as details of how the vendors plan to deliver the goods or services. Evaluation of the vendors is then conducted using a set of criteria that can include relevant previous track records, referrals from past clients, quality records, comprehensiveness of the proposal, etc. The proposals should also be checked for conformity with existing organizational policies. For example, if the client organization specifically prohibits the use of bricks because bricklaying is not productive, then the proposals should be checked to ensure that this prohibition is not breached.

b. Completing vendor selection

When evaluating the bid prices submitted by vendors, the project team should also build up its own independent estimates in-house to predict what the costs of the goods procured are likely to be. The comparison can then take place to determine if there is a significant difference between the in-house estimates and the bid price submitted by a vendor. In the construction industry, the in-house estimates are typically provided by the consultant quantity surveyor. Essentially, the comparison is made between the estimates of the consultant quantity surveyor and the contractor's bid price. If the difference is not significant, then there is some assurance that the contractor's bid price is not unreasonable.

However, if the difference is significant, further investigations should be made to determine what causes this large discrepancy. It is also necessary to screen potential vendors to remove them from the lists of vendors if they do not meet the minimum requirements relating to relevant experience, track records, etc. Vendors who have been blacklisted recently by the authorities or by the client organization should also be removed from the lists used by the project team to shortlist potential vendors. Checks should also be made with various rating agencies where past records of potential vendors may be available for viewing. These can include checks on quality performance if the national agency does maintain a listing of projects completed by various vendors and their respective performance. The financial standing of potential vendors may also be checked through a reliable credit rating agency.

Where such information is not readily available, the project team can consult the expert with the domain knowledge in the subject matter or who have had experience working with a particular vendor in the past. The information thus collated can be used to evaluate the appropriateness and competitiveness of vendors using a weighting system. In a situation where there is only one vendor, the project team can enter into a discussion with the vendor to negotiate the contract. This is where both parties seek to determine a fair and reasonable price through negotiation that must be transparent and above board to everyone concerned.

c. Outcome of vendor selection

After a vendor has been selected, a formal contract should be entered into by the buyer and the seller. The formal contract is a legally binding agreement that spells out the obligations of both parties. It essentially sets out the scenario where the seller provides the goods or services with the buyer paying for them. The contract signed by both parties is enforceable in the courts of law. If there is a breach of contract by either party, the aggrieved party can sue for damages. The contract needs to be signed by persons authorized and empowered to do so on behalf of their organizations, including matters relating to payments stipulated in the contract. Where prior arrangements have been made, all such contracts may also be routed through a central procurement office at headquarters. Otherwise, a decentralized procurement approach at the project level may be adopted with a contract manager or contract administrator handling all contractual matters related to procurement.

Contract administration plays an important role to ensure that the vendor delivers on the agreement and comply with all the terms and conditions that both parties have agreed to. In large organizations, there may also be in-house legal counsels to advise on contractual and legal matters relating to procurement. Legal recourse may be pursued if there is a breach of contract. The contract administrator also works with the vendors and contractors to coordinate their schedules, deliveries and performance. Proper co-ordination is necessary to ensure that the work of one vendor does not adversely affect the performance of another vendor. This is where project integration management is expected to feature strongly with clear communications taking place between different stakeholders. At the same time, contract administration also monitors performance and tracks quality deliverables as the project progresses. This is necessary for accepting the work done by the vendors and to approve payments.

Three inputs are needed for administering procurement contracts. Firstly, performance reports on the vendor's performance should be prepared. These reports can be prepared by the project team or the consultant quantity surveyor who visits the site. Secondly, such visits determine if the results of the work have been delivered correctly, of the right quantities and if the specified quality standards have been met. Performance reports and inspections of work results provide the basis for making payments to the vendors. Thirdly, administering procurement contracts also needs to deal with change requests. When changes are requested, contractual amendments may have to be made or a new contract created to account for the additional work or the change in the nature of work. If approved, variation orders are also issued to instruct and document the change. Nevertheless, if the change is significant and deviates substantially from the original intent of the agreement, the buyer and vendor may disagree to the change. Such a situation may lead to disputes, claims and adversarial actions in extreme cases.

12.8 Completion of Contract Administration

a. What completion involves

After the vendor has fulfilled his part of the agreement and delivered the goods and services, performance reviews and audits are then conducted by the buyer. The purpose of the reviews and audits is to ensure that the vendor has indeed delivered and complied with everything that have been agreed in the contract. For example, this can include testing and commissioning of the fire protection system installed in the building. The outcome of the reviews and audits is the performance report on the vendor. Payment is made to the vendor upon receipt of a satisfactory performance report. The payment procedure should follow that stipulated in the contract which spells out how and when payment requests are to be submitted. Thereafter, payments have to be made within the timeframe stipulated in the contract or within a reasonable time in exceptional cases. Being the client, a buyer may also specify how and when the invoices are to be paid taking into account the normal credit terms of 30 days practised in the construction industry. If the buyer is a long-term valued client, he may also stand to benefit more discounts offered by the vendor.

The completion of contract administration should also account for approved changes that occurred. By tapping on a procurement contract change control system, all documents associated with the changes are tracked and recorded. The project team or central procurement office should therefore put in place a user friendly documents management system to facilitate the search and easy retrieval of the relevant records if a dispute arises. This is for the purpose of anticipating and managing disputes that may occur in the future. Administering claims is also part of the completion process for contract administration for procurement. Such claims may arise because of disagreement over a change, apportionment of costs arising from a change and contesting who should pay for the change.

b. Contract closure and close-out

The involvement of the project team, client, vendor and relevant stakeholder is necessary to finalize product verification to confirm that the contract has indeed been completed fully. This is for the purpose of contract closure and can take the form of a site walk-about and inspection of the works installed. In a worst case scenario, closure may also come about because a contract has been terminated due to non-performance. For example, the services of a recalcitrant builder who persistently delivered poor quality standards may be terminated. Closure may also happen when a builder becomes insolvent. In all cases of closure, whether due to successful completion or termination, all documents relating to the procurement contract should be formally signed, updated and recorded. Reasons for the closure such as acceptance of the work or termination should be indicated in the records for future reference. A close-out follows the contract closure.

The close-out exercise should include a contract file comprising of a complete set of procurement records that is properly indexed for easy retrieval. The records should include all the pertinent financial information relating to the procurement contracts, information on how the vendors have performed and whether the works procured have been accepted. Such records become part of the historical assets of the organization that can be referred to by the project team for future similar projects. At the same time when close-out takes place, formal closure in the form of an official letter should be sent by the contract administrator to the vendor. Apart from informing the vendor that his work is acceptable and that the contract is considered closed, it is also basic courtesy to thank the vendor for the services rendered and for the opportunities to work together again if possible in the future.

12.9   Revision Questions

1. What are some of the resources that a typical building project need to procure?
2. What does the process of project procurement management involve?
3. What questions need to be asked for procurement planning?
4. Why is an assessment of marketplace conditions necessary for procurement planning?
5. What are the reasons for procurement planning to make reference to the scope statement and work breakdown structure?
6. How are timings for various purchases determined in the project management plan?
7. What are "make" or "buy" decisions when procurement planning is completed?
8. When should "make" decisions be made?
9. When should "buy" decisions be made?
10. What are some of the general principles for contracts used in procurement?
11. What is a fixed price contract?
12. What are cost reimbursable contracts?
13. What are direct costs and indirect costs in cost reimbursable contracts?
14. Why is the buyer assuming the highest risk in the cost plus percentage contract?
15. Why is the buyer assuming high risk in the cost plus fixed fees contract?
16. Why is the seller assuming low risk in the cost plus incentive fees contract?
17. Why is the seller assuming high risk in the fixed price contract?
18. What is a "Not to exceed" clause in procurement contracts?
19. Why is it necessary for "Time and materials contracts" to include a schedule of rates?
20. What is a term contract?
21. What does the procurement management plan cover?
22. Why is it necessary to provide a description of work for procurement purposes?
23. What are some of the preparation works needed for contracting?
24. What are some of the key documents needed to organize contracting materials for procurement purposes?

25. Is there a difference between a bid and an Invitation for Bid?
26. Is there a difference between a quotation and a Request for Quote?
27. Is there a difference between a proposal and a Request for Proposal?
28. Why is it important to have a set of criteria to evaluate and select vendors?
29. Why is it necessary to update the description of work during the procurement process?
30. How does pre-qualification help in the procurement process?
31. How does potential vendors benefit from bidder conferences?
32. What need to be done to complete the vendor selection process?
33. What take place after a vendor has been selected?
34. What is the role of contract administration in procurement management?
35. Why are the three inputs of performance reports, work results and change requests important to contract administration?
36. What take place prior to completing contract administration?
37. What are payment requests and credit terms?
38. What does contract close-out entail?

Chapter 13
Project Ethics and Professional Conduct

13.1 How Is This Relevant for Project Management?

a. Things that can go wrong

The practice of project management involves many stakeholders who often times take care only of their self-interests to the detriment of others. Granted that project management straddles nine broad knowledge areas, it is inevitable that ethical practices and professional conduct may be compromised by some stakeholders to safeguard their own organizational survival and profitability. A stakeholder may misrepresent the project scope in order to give his firm an advantage over others in the technological options that are available only through his organization. In cost management, a stakeholder may unreasonably withhold payments to vendors in order to facilitate his own cash flow; although this is now rendered more difficult for him to do so through the enactment of the Security of Payment Act. In project time management, Parkinson's Law may be present when employees deliberately delay the completion of the tasks assigned to them; i.e. work expand to fill the time available for its completion.

Because of inadequate skills training for workers, poor workmanship standards provided by a stakeholder can undermine good project quality management. Conflicts of interests can also surface in human resource management when a project manager only hires his own people who do not meet the minimum employment requirements. In projects that are behind schedule, stakeholders may recklessly rush to complete tasks without seriously assessing the associated risks involved, thereby causing accidents to happen. In project communications management, stakeholders may deliberately refrain from providing the necessary information even though they are in possession of such information. This refusal to provide the information available is to place them in a better position relative to other stakeholders.

© Springer Nature Singapore Pte Ltd. 2018
L.S. Pheng, *Project Management for the Built Environment*,
Management in the Built Environment,
https://doi.org/10.1007/978-981-10-6992-5_13

Procurement practices may not be entirely transparent because project stake-holders may be receiving commissions from vendors whose products and services have been recommended for purchase even though these may not provide the best value for money to the project. Finally in integration management, an alert stake-holder may spot an error but chooses only to take care of his own portion of work in the project without informing other stakeholders who are also equally affected by that error. Hence, these examples suggest that ethical practice and professional conduct are very real issues in project management as the various stakeholders come into contact with their counterparts across a wide range of tasks over time. While some of the decisions or omissions may not be illegal, these are still lapses in judgement on the part of the project stakeholders. Honesty means doing the right thing even when no one is watching you.

13.2 What Constitutes Good Ethical Practice?

a. When in doubt, don't

In the course of working on a project that spans over a long duration, the project manager is inevitably faced with dilemmas that require him to make hard decisions. Some of these decisions are to be taken when little or no information is available. Faced with the pressure of time and lack of information, a project manager may be tempted to take the easy way out to make a decision that is not anchored on objectivity and integrity. The ready response when faced with such a situation is for the project manager to not do anything rash when there is doubt. At all times, even when there is the slightest of doubt, the project manager must always err on the safe side and exercise caution in all that he does. He must maintain perseverance in seeking the information that he needs to make good decisions and not settle for anything lesser.

The project manager should not succumb nor crumble even under intense pressure from all sides. Instead, he should see this as an opportune time for him to show his leadership capabilities to his project team members. No problems or difficulties last forever. Like the dark night that passes, dawn will soon arrive. This should always be the positive mind-set of the project manager when he encounters doubtful and difficult situations. In making decisions, the project manager should always ensure there is no room for misinterpretation or misconception by other stakeholders. He should always avoid errors of judgement and refrain from shady or wrong actions that may be construed by others to give rise to conflicts of interests or impropriety. In the final analysis, the project manager should discharge his pro-fessional responsibilities in the following five areas.

Firstly, the project manager must always ensure integrity in all that he does and decides. This means being honest and fair in all his dealings. Secondly, as a member of the wider professional community, the project manager should also play a part in contributing to existing knowledge in the theories and practice of project management. This means that when called upon, the project manager should be ready to share his views and knowledge to students and fellow project managers. Thirdly, through such interactions, the project manager in turn gains further insights to enhance his own know-how to apply his knowledge professionally. Fourthly, the project manager needs to ensure that all stakeholder interests are balanced and met as far as is possible. This means that the interests of one single stakeholder should not jeopardize the interests of others. For example, the project manager needs to ensure that speeding up project completion should not compromise the safety of other stakeholder especially the employees tasked with excessive overtime work.

Finally, the project manager should respect differences in beliefs and opinions. This is especially so in modern-day projects that is increasingly globalized in nature. The project manager should be sensitive to religious beliefs and cultural differences. Apart from respect, this means that the project manager should also make special efforts to integrate differences among stakeholders for them to contribute collectively to the overall project goals. Attempts by insensitive stakeholders to make fun of certain cultural norms or religious beliefs must not be condoned at all times. The project manager must always respect the laws, people, religions, cultural beliefs, traditions, practices and values of the country he works in. A conscious attempt must be made to avoid ethnocentrism in which a person believes that his own culture, values and beliefs are better than those of others.

13.3   Industry Standards and Regulations

a. Guidance for the professionals

As part of the wider community of project management professionals, the project manager is very likely to be a member of a related professional institution with its own standards and regulations that govern membership. Most professional institutions generally have codes of conduct that members need to adhere to or risk disciplinary actions or in the worst case scenario debarment. Professionals are therefore expected to comply with an honourable set of principles and policies governing high standards of conduct and behaviour. In addition, the professional is expected to contribute to the advancement of the discipline he is in. This can be in the form of mentoring younger members, giving talks, collaborating in research, engaging with communities and serving in technical committees. In the process, the professional also learns to upgrade himself continuously. In all these activities, the professional is expected to be honest in all his dealings.

The test for honesty rests on whether another professional peer would reasonably exercise the same judgement, make the same decision and take the same action. To avoid any doubts, the professional must make his position clear to clients or other stakeholders who may be affected by his actions. Situations where such doubts may arise include transactions where the professional may be perceived to have an unfair advantage over others and from which he is able to profit from the project. Any semblance of a conflict of interests should be highlighted immediately to clients and employers. Even if the situation is such that there is actually no conflict of interests, but there are hints of impropriety, the professional should disclose this quickly to avoid further misunderstanding. All dealings must be seen to be above board and conducted at arms-length.

b. Industry regulations

Apart from adherence to codes of conduct through their membership of professional institutions, professionals are also guided by other standards and regulations that govern practices in a specific industry. The construction industry, for example, has definitive standards, regulations and codes of practice which building professionals need to be aware of and adhere to in the course of their work. Some industry standards are recommended good practice which building professionals should follow. An example is the standard for the implementation of business continuity management in organizations. In this case, the building professional can choose not to implement business continuity management for his small business outfit and hence is not affected by the standard.

On the other hand, regulations and statutory requirements set out the framework within which building professionals are expected to follow and must put in practice. Hence, building professionals need to be familiar with, for example, provisions of the Workplace Safety and Health Regulations and how these impinge on their work and sphere of influence. Over time, these standards and regulations may change. It is therefore important for building professionals to continuously learn and upgrade their knowledge. This is readily facilitated by attending continuing professional development programs. Being a member of a relevant professional institution is also helpful for this purpose because new requirements will be disseminated to members for their update.

c. Intellectual property

In the course of working, the building professional collaborates with others on various tasks in the project. Information is shared and received. Some of the information is created by others for specific purpose, but shared among stakeholders in the project team. When this occurs, the building professional must at all times respect the property or information that belongs to others. Recognizing the intellectual work of others and honouring intellectual property rights must always be a foremost consideration. The building professional must always give credit where credit is due in areas as diverse as designs, models and information datasets. Patents, copyrights and registered trademarks belonging to others must be acknowledged at all times.

13.4 Forging Ahead

a. Other related issues

The building professional owes a duty of care to clients who can be from both the private and public sectors. In discharging his responsibilities, the building professional must exercise accountability and fairness in all his dealings in both sectors, especially where public funds are being used. There must not be any compromise even if it appears that generous public funds are readily available for the project. A conscious attempt must be made to avoid wasteful spending of taxpayer monies through over-specifying or use of ultra-luxurious products. A good balance should always be maintained to ensure that public health and safety is not compromised even in the face of pressure by clients to cut costs. In the course of his work, the building professional must always remain honest in all dealings and to enforce the truth.

This is especially important in advertisements where unfounded claims must not be made. Press releases and public announcements must always reveal only the truth even if the truth hurts. This must always be the practice even if telling the truth may place the organization in a difficult situation such as in the aftermath of disastrous project outcomes or accidents. Similarly, when speaking at a public forum, the building professional should not make unfounded claims or allegations that cannot be verified. He should speak honestly and truthfully on the subject matter and to the best of his knowledge at that point in time. At all times, the building professional must exercise discipline to ensure that confidential information is kept confidential and that the privacy of others is not breached. Information that is sensitive should never be disclosed. Non-disclosure agreements must be honoured at all times.

b. Gifts and briberies

The presentation of gifts is a frequent occurrence to build relationships and to show appreciation to someone. The acceptance of gifts may however be construed to be inappropriate in some instances especially when the gifts concerned are luxurious items or are costly. The professional should exercise restraint not to accept such gifts even if the act of giving is well intentioned but run contrary to established organizational policies that forbid employees from accepting gifts. In situations where rejecting the gifts may lead to embarrassment on the part of the giver, the professional should check with his organization to determine if such gifts can be accepted and if so, under what circumstances. The onus is on the professional to inform his employer of the gifts for this to be made transparent and above board.

At all times, the professional must not accept any gifts and payments given in return for favours or to influence project outcomes. It should however be noted that there can be exceptions where the nature of some gifts is concerned. If the gift is given as part of the established traditions, customs or culture of a country in which the professional is working in, the acceptance of such a gift is considered to be normal. Nevertheless, if the gift is exceptionally luxurious and outside of the norms, the professional should not accept it. This is especially so when there are established policies within the organization that prohibit employees from accepting gifts that exceed a stipulated amount. If the value of the gift exceeds this stipulated amount, the professional must not accept the gift because doing so is certain to lead to a conflict of interest with his organization. Some gifts can be luxurious and include paid cruise holidays, boxes of cigars, crates of fine wine and meals at expensive restaurants. Other smaller gifts can include appreciation medals presented to project team members.

The professional should therefore be aware of the nature of different gifts and to make decisions to accept or refuse such gifts in line with established company policies. Similarly, on the other side of the fence, the professional may also present gifts to others. Again, this act of giving must always be in line with established company policies. Gifts or payments to others to buy and gain their influence must not be condoned. This is true even in countries where gifts and payments are expected as part of the culture and is seen as a tradition rather than as kick-back, corruption or bribery. When faced with such a situation, it is to the interest of the professional to refrain from doing so. Even if such payments are normal in the host countries which do not consider such transactions as being illegal, the professional can still be charged for corrupt practice if he returns to his home country which does not allow illicit payments and bribes.

c. Doing the right thing

In the course of our work, we are often hard pressed to make decisions which on the surface do not seem to be wrong. This perception arises because no one gets hurt or the company did not suffer any real monetary loss and no one finds out what you have done. Take the following hypothetical example. You are the project manager of a building site and have access to the site office safe which contains $10,000 in petty cash. Today is Friday and you have had an extremely busy and tiring day at work. At 5 pm, you suddenly remember that you have an appointment with a tour agency who has promised you an attractive discount for a family travel package provided you pay a deposit of $5000 to them by 7 pm.

Your bank is already closed by now and the daily withdrawal limit if you use the Automated Teller Machine is $2000. For convenience, you contemplate dipping your hand into the site office safe to "borrow" $5000 from the petty cash amount. This is not your money. It belongs to your employer. Nevertheless, you think no one will know since everyone has already left the site office. You have every intention of returning the money in the safe when your bank reopens on the following Monday. No one is watching. Nobody will ever know. And you get to keep your attractive discount promised by the tour agency. Is this the right thing to do?

13.5 Discussion Questions

1. Why are ethical practices and professional conduct matters of the heart?
2. Why should the project manager always take the high road?
3. When in doubt over certain ethical and professional conduct issues, what should the project manager do?
4. How can the project manager ensure integrity?
5. What can the project manager do to contribute to the knowledge base?
6. Why must differences in belief systems be respected by project team members?
7. Why is it crucial to apply professional knowledge?
8. Why must the project manager balance stakeholders' interests at all times?
9. What should the project manager do if any appearances of impropriety have been detected?
10. Should all recommended industry standards be enforced in projects?
11. Is it alright to plead ignorance of regulatory requirements?
12. Why must credit be given when credit is due?
13. What are some matters that the project manager must keep confidential?
14. Why is it necessary to keep all advertisements and press releases truthful?
15. When may the project manager accept gifts?
16. What should the project manager be respectful of in international construction projects?
17. Why is ethnocentrism undesirable in the project setting?
18. Upon successful project completion in a country in the Middle East, is it alright for the project manager to accept a Ferrari sports car as a gift from the client as this is a common practice in that country?
19. As the project manager, is it alright for your former university classmate who works for a vendor in your current project to buy you lunch at a five-star hotel?
20. Should the project manager condone whistle-blowing?

Bibliography

1. Low, S.P., and W.M. Alfelor. 2000. Cross-cultural influences on quality management systems: Two case studies. *Work Study* 49 (4): 134–144.
2. Low, S.P., and C.H.Y. Leong. 2000. Cross cultural project management for international construction in China. *International Journal of Project Management* 18 (5): 307–316.
3. Low, S.P., and K.O. Lim. 2000. Success factors for design-and-build in civil engineering projects: Two case studies. *Bulletin of the Institution of Engineer, Malaysia* (August Issue): 57–64.
4. Low, S.P., and C.H.Y. Leong. 2001. Asian management style versus western management theories: A Singapore case study in construction project management. *Journal of Managerial Psychology (Special issue on Asian Management Style)* 16 (2): 127–141.
5. Low, S.P., and D. Wee. 2001. Impact of ISO 9000 on the reduction of building defects. *Architectural Science Review* 44 (4): 367–377.
6. Low, S.P., and E.T.W. Fong. 2002. Preparations for ISO 9001: 2000—a study of ISO 9000: 1994 certified construction firms. *Construction Management and Economics* 20 (5): 403–413.
7. Low, S.P. 2002. ISO 9000 quality management systems for construction safety. In *Building in value*, ed. R. Best, and G. de Valence, 354–372. London: Butterworth-Heinemann.
8. Low, S.P., and Y.P. Chin. 2003. Integrating ISO 9001 and OHSAS 18001 for construction. *Journal of Construction Engineering and Management* 129 (3): 338–347.
9. Low, S.P., and S.P. Loh. 2003. Organizational culture and construction quality: A systemic study of contractors in Singapore. In *Joint International Symposium of CIB Working Commissions on Knowledge Construction*, ed. G. Ofori and F. Ling, Singapore: National University of Singapore; and CIB, Joint International Symposium of CIB W55, W65 and W107 on "Knowledge Construction", 567–578, 22–24 Oct.
10. Low, S.P., and J.A. Teo. 2004. Implementing TQM in construction firms. *Journal of Management in Engineering* 20 (1): 8–15.
11. Low, S.P., and S.H. Mok. 2004. Implementing and applying Six Sigma in construction. *Journal of Construction Engineering and Management* 130 (4): 482–489.
12. Low, S.P., and H.L. Pan. 2004. Critical linkage factors between management and supervisory staff for ISO 9001: 2000 quality management systems in construction. *9th International Conference on ISO 9000 and TQM (9–ICIT)*, TQM Best Practices, ed. S. Ho and P. Suttiprasit, Hang Seng School of Commerce and Sukhothai Thammathirat Open University, 142–147, Apr 5–7, Siam City Hotel, Bangkok, Thailand.
13. Low, S.P., and S.H. Hong. 2005. Strategic quality management for the construction industry. *The TQM Magazine* 17 (1): 35–53.
14. Low, S.P., and K.K. Goh. 2005. ISO 9001, ISO 14001 and OHSAS 18001 management systems. Integration, costs and benefits for construction companies. *Architectural Science Review* 48 (2): 145–151.

© Springer Nature Singapore Pte Ltd. 2018
L.S. Pheng, *Project Management for the Built Environment,*
Management in the Built Environment,
https://doi.org/10.1007/978-981-10-6992-5

15. Low, S.P., and J.H.K. Tan. 2005. Integrating ISO 9001 quality management system and ISO 14001 environmental management system for contractors. *Journal of Construction Engineering and Management* 131 (11): 1241–1244.

16. Low, S.P., and S.M. Tan. 2006. The evaluation and management of contractors' creditworthiness by suppliers of building materials in Singapore. *The Malaysian Surveyor, Institution of Surveyors Malaysia* 40 (3): 40–46.

17. Low, S.P., and T.C. Quek. 2006. Environmental factors and work performance of project managers in the construction industry. *International Journal of Project Management* 24 (1): 24–37.

18. Low, S.P. 2007. Managing building projects in ancient China. A comparison with modern-day project management principles and practices. *Journal of Management History* 13 (2): 192–210.

19. Low, S.P., J. Barber, and P.S.P. Ang. 2007. Comparative study of the Price-Quality Method in Singapore, Hong Kong, the United Kingdom and New Zealand. *International Construction Law Review* 24 (3): 318–342.

20. Low, S.P., J.Y. Liu, and Q.S. He. 2008. Management of external risks: Case study of a Chinese construction firm at infancy stage in Singapore. *International Journal of Construction Management* 8 (2): 1–15.

21. Low, S.P., J.Y. Liu, and S.S. Soh. 2008. Chinese foreign workers in Singapore's construction industry. *Journal of Technology Management in China* 3 (2): 211–223.

22. Low, S.P., and S.B. Lim. 2008. Value engineering and value management: Case study of an airshow exhibition centre, *Proceedings of the 9th International Value Management Conference: Achieving Sustainable Values through Collaboration*, The Hong Kong Institute of Value Management, 29th Oct-1st Nov, Hong Kong Polytechnic University, ed. G.Q.P. Shen et al., 224–237.

23. Low, S.P., and W.H. Chia. 2009. Middle management's influence on the effectiveness of ISO 9000 Quality Management Systems in Architectural Firms. *Architectural Engineering and Design Management* 4 (1): 189–205.

24. Low, S.P., and C.F. Ong. 2009. *Quality construction in the building industry*, 25–35. Institution of Engineers, Singapore, June: The Singapore Engineer.

25. Low, S.P., J.Y. Liu, and Q.S. He. 2009. External risk management practices of Chinese construction firms in Singapore. *KSCE Journal of Civil Engineering (Korean Society of Civil Engineers)* 13 (2): 85–95.

26. Low, S.P., and J.Y. Liu. 2009. Causes of construction delays and their contractual provisions in Mainland China. *International Construction Law Review* 26 (2): 463–488.

27. Low, S.P., J.Y. Liu, and W.Y. Leow. 2010. Work-family life of consultant quantity surveyors in Singapore. *Journal of Quantity Surveying and Construction Business* 1 (1): 1–23.

28. Low, S.P., and H.I. Low. 2011. Effects of organizational behavior on the maintenance of ISO 9001 quality management systems in the construction industry, *Proceedings of the 2nd International Conference on Project and Facilities Management*, 18th–19th May 2011, Centre for Construction, Building and Urban Studies, University of Malaya, 9–29.

29. Low, S.P., and K. Seet. 2011. Enhancing construction quality through TQM. *The Singapore Engineer*, April, 16–22.

30. Low, S.P. 2011. Management of change in Singapore's Programme for Rebuilding and Improving Existing (PRIME) Schools, In *Proceedings of the 1st International Symposium on Conducive Learning Environment for Smart Schools (CLES 2011)*, ed. N. Utaberta et al., 1–23, Jan 17, organized by Universiti Kebangsaan Malaysia (UKM) and supported by the Board of Engineers (Malaysia), Cyberjaya, Kuala Lumpur, Malaysia.

31. Low, S.P., X.P. Deng, and L. Lye. 2012. Communications management for upgrading public housing projects in Singapore. *Structural Survey* 30 (1): 6–23.

32. Low, S.P., J.Y. Liu, S.H.M. Ng, and X. Liu. 2013. Enterprise risk management and the performance of local contractors in Singapore. *International Journal of Construction Management* 13 (2): 27–41.

33. Low, S.P. 2013. Sharing roles and responsibilities: Emerging trends and legislative controls influencing project delivery methods in Singapore. In *Proceedings of the 2nd International Conference of Construction Project Delivery Methods and Quality Ensuring Systems,* ed. S. Furusaka, et al., 251–261, Sept 26–28, Kyoto University.

34. Low, S.P., and J. Ong. 2014. *Project quality management. Critical success factors for buildings.* Berlin: Springer.

35. Low, S.P., S. Gao, and W.L. Tay. 2014. Comparative study of project management and critical success factors of greening new and existing buildings in Singapore. *Structural Survey* 32 (5): 413–433.

36. Low, S.P. 2015. 50 years of construction project delivery methods and quality ensuring systems in Singapore. In *Proceedings of the International Conference on Construction Project Delivery Methods and Quality Ensuring Systems,* ed. S. Furusaka et al., 67–89, Nov 19–20, Kyoto University.

37. Low, S.P. 2015. A review of construction productivity indicators in Singapore, *The Singapore Engineer,* August 2015, 24–30.

38. Low, S.P., and R. Zhu. 2016. Service quality for facilities management in hospitals. Berlin: Springer.

Printed in the United States
By Bookmasters